This book presents an expansion of the highly successful lectures given by Professor Barton at the Polytechnical Institute of Milan under the auspices of the Accademia Nazionale dei Lincei.

The book explores the invention of new chemical reactions for use in the synthesis of biologically and economically important compounds. It begins with a mechanistic study of the industrial importance of the pyrolysis of chlorinated alkanes. It continues with a theory on the biosynthesis of phenolate derived alkaloids involving phenolate radical coupling. Included in the book is a description of the work on nitrite photolysis (the Barton Reaction) which involved the invention of new radical chemistry leading to a simple synthesis of the hormone, aldosterone. In two final chapters Dr Shyamal Parekh views Professor Barton's pioneering work from the modern perspective, with a review of recent applications in industry and research.

The book should prove to be an enlightening and exciting review of over fifty years of creative chemical research.

Half a century of
free radical chemistry

Lezioni Lincee
Sponsored by *Foundazione IBM Italia*
Editor: Luigi A. Radicati di Brozolo, Scuola Normale Superiore, Pisa

This series of books arises from lectures given under the auspices of the Accademia Nazionale dei Lincei and is sponsored by *Foundazione IBM Italia.*
The lectures, given by international authorities, will range on scientific topics from mathematics and physics through to biology and economics. The books are intended for a broad audience of graduate students and faculty members, and are meant to provide a '*mise au point*' for the subject they deal with.
The symbol of the Accademia, the Lynx, is noted for its sharp sightedness; the volumes in the series will be penetrating studies of scientific topics of contemporary interest.

Already published

Chaotic Evolution and Strange Attractors: D. Ruelle
Introduction to Polymer Dynamics: P. de Gennes
The Geometry and Physics of Knots: M. Atiyah
Attractors for Semigroups and Evolution Equations: O. Ladyzhenskaya

Half a century of free radical chemistry

DEREK H. R. BARTON
in collaboration with
SHYAMAL I. PAREKH
Department of Chemistry
Texas A & M University, USA

Published by the Press Syndicate of the University of Cambridge
The Pitt Building, Trumpington Street, Cambridge CB2 1RP
40 West 20th Street, New York, NY 10011-4211, USA
10 Stamford Road, Oakleigh, Melbourne 3166, Australia

First published 1993

A catalogue record for this book is available from the British Library

Library of Congress cataloguing in publication data

Barton, Derek, Sir, 1918–
Half a century of free radical chemistry/Derek H. R. Barton in collaboration with S. I. Parekh.
p. cm. – (Lezioni Lincee)
Includes index.
ISBN 0–521–44005–X
1. Free radical reactions. 2. Organic compounds – Synthesis.
I. Parekh, S. I. (Shyamal I.) II. Title. III. Series.
QD471.B313 1992
547 – dc20 92–24681CIP

ISBN 0 521 44005 X hardback
ISBN 0 521 44580 9 paperback

Transferred to digital printing 2004

Contents

Preface

First, I must express my appreciation for the honour of being invited by the President, Prof. Giorgio Salvini, of the Accademia Nazionale dei Lincei to give the Lezioni Lincee for 1990–1991. The added responsibility of writing a short book summarizing these lessons was also stimulating. In view of the long-standing research interests of my host at the Dipartimento di Chimica of the Politecnico di Milano, Professor Francesco Minisci, a very distinguished radical chemist, I decided to talk about my 50 years of research on the chemistry of free radicals. Some of my audience will surely demand how something that happened in chemistry fifty years ago could have any relevance to the present day. I hope to show that lessons can be learnt from the past about the philosophy of chemical research. Older chemists always seem to be interested in this subject, although many also become interested in the origins of life.

I have always tried to select a problem whose solution would be significant. On the other hand, the problem must be one that could be solved with the means at one's disposal. Frequently, I have chosen problems relevant to the chemistry of natural products.

The first problem that I discuss in this book was the major part of my PhD thesis. The problem, the synthesis of vinyl chloride, was chosen for me by the circumstance of war. It was of national importance. When the problem has been chosen, one has to think about it. The thought process is based on known literature, but the creative process is to imagine the solution to the problems in terms of the unknown. The creative thinking in the first chapter is about the interpretation of what seemed a baffling problem of experimental observations. The comprehension of the meaning of the observations led to a theory of considerable predictive value.

The second chapter is initiated by thoughts about the problem of the biosynthesis of morphine. Ever since Sir Robert Robinson deduced the correct constitution for morphine in the early 1920s, he speculated about the biosynthesis. Of course, he was right in proposing that morphine was an oxidized and cyclized benzyl isoquinoline alkaloid. However, he and later Schöpf were wrong in assuming that the structure of Pummerer's ketone was correct. Anyone who read Pummerer's paper critically would have seen that it was wrong. Anyone? In fact, the wrong formula was accepted for 30 years before someone (I) saw what the true constitution should be. Then, logic led from the right formula for Pummerer's ketone to right biosynthetic proposals for morphine and many other phenolic alkaloids.

The third chapter concerns the synthesis of aldosterone and its congeners. The problem of synthesis arose because of the scarcity of the hormone and a pressing need to study its biological effects. An elegant solution to the problem required the invention of a new reaction. So the creative thought was in thinking what that might be; and it was.

The fourth chapter concerns a family of radical reactions which were invented to solve the problem of how to remove secondary hydroxyl groups in complicated amino-glycoside antibiotics. This biologically important problem was solved by the invention of a new reaction. It changed carbohydrate chemistry and introduced many chemists to high yielding radical reactions for the first time.

The fifth chapter has been mainly written by Dr Shyamal I. Parekh, who has also been responsible for all the production process in preparing the final manuscript. The general philosophy is again apparent. A new reaction was needed to manipulate the carboxyl group in peptides and in compounds of the arachidonic acid cascade, and it was invented for the purpose.

The final chapter, written by Dr Parekh, is a summary of recent applications by other groups of the reactions invented and discussed in Chapters 4 and 5.

I take this occasion to thank Prof. Minisci and his colleagues for their kindness in making my stay in Milano so enjoyable and initiating the project of writing this book.

D. H. R. Barton
Milano, June 17th, 1991

1

The pyrolysis of chlorinated hydrocarbons

I was admitted to Imperial College in October 1938 when I was 20 years old. Because of the good marks I had obtained in the entrance examinations I was allowed to proceed directly into the second year. I completed the BSc with First Class Honors in 1940. Because of what turned out to be a minor problem with my heart, I had earlier been called to the Army and rejected. In any case, I was in a reserved profession and expected to work on a subject of national importance.

So, in 1940 I began my PhD work which mainly involved the synthesis of economically important vinyl chloride. The project was financed by the Distillers Company and I worked in association with a German refugee chemist, Dr M. Mugdan. We started off by studying heterogeneous catalysts for the addition of hydrogen chloride to acetylene.[1] At that time this was the preferred industrial process for vinyl chloride. Then, early in 1941, I began to study the homogeneous, non-catalysed

$$\underset{\mathbf{1}}{Cl\text{-}CH_2\text{-}CH_2\text{-}Cl} \xrightarrow{\Delta} \underset{\mathbf{2}}{CH_2{=}CHCl} + HCl$$

pyrolysis of ethylene dichloride **1** into vinyl chloride **2** and hydrogen chloride, the method almost universally used at the present time. This deceptively simple reaction turned out to be unusually complex.[2-4] It provides a good example of radical chemistry and eventually led to a correlation between structure and mechanism.

Many organic molecules decompose in the gas phase by the

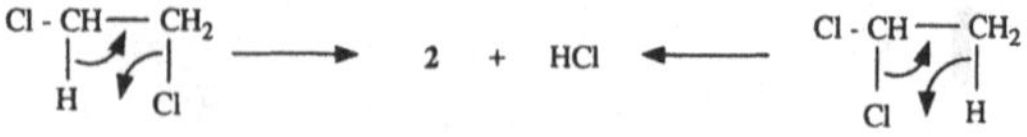

$$\text{Cl-CH—CH}_2 \text{ (H, Cl; } \mathbf{1}) \longrightarrow \mathbf{2} + \text{HCl} \longleftarrow \text{Cl-CH—CH}_2 \text{ (Cl, H; } \mathbf{3})$$

Scheme 1

so-called 'unimolecular' mechanism. The transition state is depicted in Scheme 1, and such a mechanism should also be available to 1,1-dichloroethane **3**.

The gas phase pyrolysis of 1,1- and 1,2-dichloroethane affords nearly quantitatively HCl and **2**. For my PhD work in 1940–1942 I studied these two reactions in a glass tubular-flow reactor. When the glass is clean, there is always a fast heterogeneous reaction. However, the active centers of the surface are soon poisoned and the reaction then becomes homogeneous. This is demonstrated by packing the reactor with glass to vary the surface area to volume ratio. After the usual 'poisoning' period, the packed reactor gives the same reaction rate as the unpacked reactor. Hence, the final reaction is not heterogeneous.

For several weeks these two dichlorides decomposed at the same rate by the unimolecular mechanism. But one day, without warning, the ethylene dichloride started to decompose much faster than the 1,1-dichloroethane, such that I could obtain the same conversion at 100 °C lower temperature. After reflection, I realized that the ethylene dichloride used had been recovered from the dry ice trap and then redistilled before use. Normally the 1,1- and 1,2-dichloroethanes were purified by careful fractional distillation. Clearly, my recovered sample contained a catalyst, or lacked an inhibitor. The latter seemed more probable, so I treated the 1,2-dichloroethane with chromic acid or potassium permanganate, shaking overnight. After redistillation, the purified dichloride still gave variable results, some days decomposing very fast and some not. The simply distilled material always had a constant rate of decomposition. The rate for 1,1-dichloroethane was always constant and independent of comparable chemical treatment.

Finally, I identified another factor, a variable air leak. When I eliminated this, both dichloroethanes decomposed slowly by the unimolecular mechanism. When I let in a controlled flow of air (or chlorine), the 1,2-dichloroethane (in contrast to the 1,1-isomer) now decomposed rapidly at a much lower temperature. I had discovered my first new reaction: the radical chain decomposition of the dichloride as in Scheme 2.

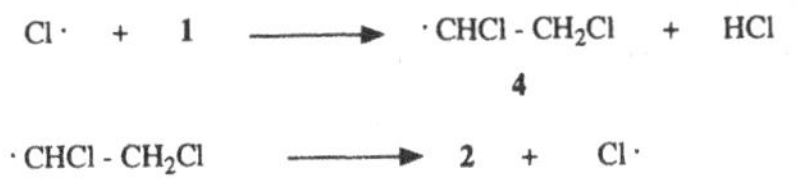

$$Cl\cdot + \mathbf{1} \longrightarrow \cdot CHCl-CH_2Cl\ (\mathbf{4}) + HCl$$

$$\cdot CHCl-CH_2Cl \longrightarrow \mathbf{2} + Cl\cdot$$

Scheme 2

The inhibitor in commercial ethylene dichloride of the 1940 era was ethylene chlorohydrin. The two form an azeotrope and so all my laborious efforts at purification by distillation failed.

Why is ethylene chlorohydrin an inhibitor of the chain (involving radical **4**) depicted in Scheme 2? This must be

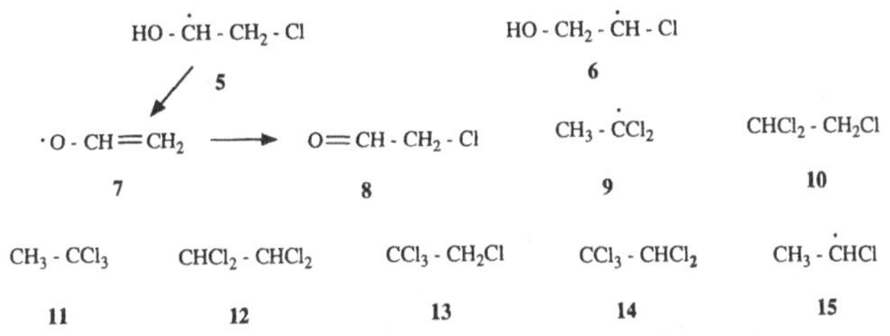

$HO-\dot{C}H-CH_2-Cl$ (**5**) $\swarrow$ $\cdot O-CH{=}CH_2$ (**7**) $\longrightarrow$ $O{=}CH-CH_2-Cl$ (**8**)

$HO-CH_2-\dot{C}H-Cl$ (**6**)

$CH_3-\dot{C}Cl_2$ (**9**) $\quad CHCl_2-CH_2Cl$ (**10**)

CH_3-CCl_3 (**11**) $\quad CHCl_2-CHCl_2$ (**12**) $\quad CCl_3-CH_2Cl$ (**13**) $\quad CCl_3-CHCl_2$ (**14**) $\quad CH_3-\dot{C}HCl$ (**15**)

because of the formation of either, or both, of the radicals **5** and **6**. The radical **5** is of a type well known[5,6] to eliminate chloride ion and a proton in solution phase to furnish radical **7** which in return would react with $Cl^\cdot$ to give **8**. Of course the radical **6** will not eliminate an $OH^\cdot$ radical (C–OH bond too strong) so it would also inhibit the chain.

1,1-Dichloroethane **3** always decomposes by the unimolecular mechanism even when oxygen, chlorine, or other radical generators are added. This is because radical attack on **3** gives the derived radical **9** which cannot carry the chain.

All this was later put on a sound basis as a result of more precise measurements of rate constants and of activation energies. However, it did not require precise measurements to predict which chlorinated hydrocarbons would decompose by a radical chain mechanism and which by the unimolecular mechanism. Clearly, if the chlorinated hydrocarbon, or the product from the pyrolysis of the chlorinated hydrocarbon reacted with chlorine atoms to break the chain then the chain mechanism would not exist. Such chlorinated hydrocarbons would decompose by the unimolecular mechanism. Monochlorinated derivatives of propane, butane, cyclohexane, etc. would afford propylene, butenes, cyclohexene, etc. All these olefins are inhibitors of chlorine radical chain reactions because of the attack of chlorine atoms at their allylic positions to give the corresponding stabilized allylic radicals which do not carry the chain.

Chlorinated ethanes could be divided into two types, those that could carry the chlorine atom chain and those that could not. 1,2-Dichloroethane **1**, 1,1,2-trichloroethane **10**, 1,1,1-trichloroethane **11**, 1,1,2,2-tetrachloroethane **12**, 1,1,1,2-tetrachloroethane **13**, and 1,1,1,2,2-pentachloroethane **14** all decomposed with enhanced rates by a chlorine atom chain mechanism. Ethyl chloride and 1,1-dichloroethane **4** did not. The reason for the latter has been explained. Ethyl chloride gave likewise the radical **15** which could not carry the chain. In 1949, in a paper with the late Professor P. F. Onyon,[7] the observations made up to that time were correlated and a number of predictions were made (Table 1). In later work all the predictions were shown to be true.

The work on the pyrolysis of chlorinated hydrocarbons, especially the catalyzed synthesis of vinyl chloride, was patented by the Distillers Company and subsequently sold to the Dow Chemical Corp. Perhaps I justified my research career within the first few months! I have never met anyone who could, or would, tell me if ethylene dichloride pyrolysis, which

Table 1.

Chloro compound		Predicted homogeneous first-order mechanism: Unimolecular	Radical chain
Ethyl chloride		x (already observed)	
1,1-Dichloroethane	**3**	x (already observed)	
1,2-Dichloroethane	**1**		x (already observed)
1,1,1-Trichloroethane	**11**		x
1,1,2-Trichloroethane	**10**		x
1,1,2,2-Tetrachloroethane	**12**		x (already observed)
1,1,1,2-Tetrachloroethane	**13**		x (already observed)
Pentachloroethane	**14**		x
1-Chloropropane		x	
2-Chloropropane		x (already observed)	
1,2-Dichloropropane		x	
1,2,3-Trichloropropane			x
n-Butyl chloride		x	
tert-Butyl chloride		x (already observed)	
2,2-Dichloroethyl ether			Possibly
β-Chloroethylbenzine			Possibly
α-Chloroethylbenzene		x	

is carried out on an enormous scale, is catalyzed by air or chlorine or not.

The pyrolysis of (say) cyclohexyl chloride by the unimolecular mechanism must involve *syn*-elimination (**16** → **17**) because the product is a *cis*-olefin. When I came across the pyrolysis of steroidal esters[8,9] for the synthesis of olefins it seemed to me that *syn*-elimination (unimolecular mechanism)

Cl

H

16 **17**

must also be involved. At that time the distinction between *syn*- and *anti*-elimination in the synthesis of cyclic olefins was not established. An analysis of the literature[10] showed the relationship between mechanism (*cis*-elimination) and stereochemistry. Thus a mechanistic study of the pyrolysis of chlorinated hydrocarbons eventually made significant contributions to the stereochemistry of elimination reactions in complex natural products.

References

1. Barton, D. H. R. & Mugdan, M. *J. Chem. Soc. Ind.* (1950), **69**, 75.
2. Barton, D. H. R. *J. Chem. Soc.* (1949), 148.
3. Barton, D. H. R. & Howlett, K. E. *J. Chem. Soc.* (1949), 155.
4. Barton, D. H. R. & Howlett, K. E. *J. Chem. Soc.* (1949), 165.
5. Buley, A. L., Norman, R. O. C. & Pritchett, R. J. *J. Chem. Soc. B* (1966), 849.
6. Edge, D. J., Gilbert, B. C., Norman, R. O. C. & West, P. R. *J. Chem. Soc. B* (1971), 189.
7. Barton, D. H. R. & Onyon, P. F. *Trans. Faraday Soc.* (1949), **45**, 725.
8. Barton, D. H. R. & Rosenfelder, W. J. *Nature* (1949), **164**, 316.
9. Barton, D. H. R. & Rosenfelder, W. J. *J. Chem. Soc.* (1949), 2459.
10. Barton, D. H. R. *J. Chem. Soc.* (1949), 2174.

2

Phenolate radical coupling in synthesis and biosynthesis; Pummerer's ketone

The pioneer in phenolate radical coupling was Pummerer. In 1925 he showed[1] that one electron oxidation of *p*-cresol using potassium ferricyanide afforded a nicely crystalline ketonic dimer of the radical in up to 25% yield. Pummerer's ketone, as it became known, was considered to result from the coupling of two *p*-cresol radicals to give the dienone **1**. This then underwent spontaneous cyclization to furnish **2**. As proof of the structure

O.
O
O
O
1
O
Ac
O
Ac
Ac_2O / H^+
O
O
3
2

2, Pummerer showed that treatment with acid and acetic anhydride afforded the diphenol diacetate **3** which was identical with the same compound made by total synthesis. Although I could see, perhaps, some driving force in the rearrangement of **1** to **2**, I could not imagine a mechanism which would permit this at room temperature in the presence of only ferro- and ferricyanides as reagents. However, structure **2** was accepted as true for 30 years and was used by Sir Robert Robinson and Prof. C. Schöpf as a model for the biosynthesis of morphine and sinomenine. It is interesting to reflect why no one before had questioned the structure given by Pummerer. It is possible that Robinson was too busy to look into past literature. Around 1952, it seemed to me that there was a much better way in which to formulate the coupling of two *p*-cresolate radicals to give a different dienone **4** as an intermediate. This only had to add the phenolate anion to the dienone to reach a very acceptable formula **5** for Pummerer's ketone. The formation of **3** would then be the result of an acid catalyzed dienone–phenol rearrangement, a well-known reaction. Still, it was not well known until the deduction by R. D. Haworth of the correct structure for santonin,[2,3] where this rearrangement had complicated the task of structural determination. However, I know that R. D. Haworth, who knew well Pummerer's ketone, did not jump the gap and question the formula **2**.

As soon as I was convinced that I was right, I quickly devised the degradation of **5** to 1-methylcyclohexane-1-carboxylic acid **6**.[4] The formation of this degradation product proved the identity of the original carbon–carbon bond coupling process.

The importance of the revised structure **5** for biosynthesis was impressive. We decided to carry out a synthesis of the important lichen compound usnic acid **7**. Although we,[5] and others,[6] were by then reconciled to the correctness of **1**, a biogenetic synthesis was welcome. Methylphloracetophenone **8** was oxidized with potassium ferricyanide to give the derived phenolate radical. These radicals coupled together in the desired manner to give usnic acid hydrate **9**, in which all the carbon skeleton had been correctly assembled. Dehydration gave usnic acid, although the overall yield was modest (15% unoptimized). This was an elegant synthesis.[7,8] Later it was proved that usnic acid was indeed biosynthesized by this route.[9,10]

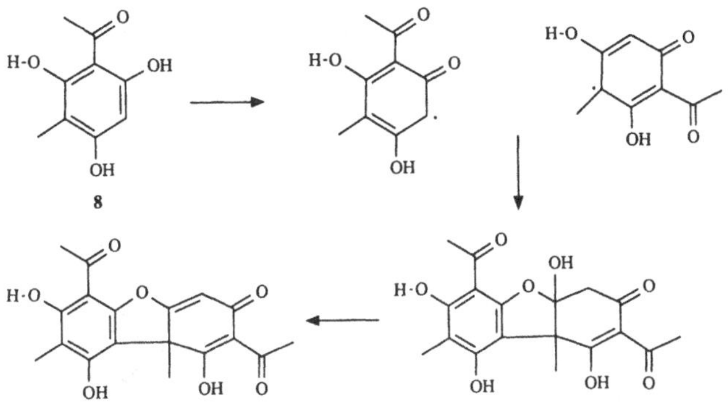

Through accepting the wrong formula for Pummerer's ketone, all the proposals by Sir Robert Robinson on the biosynthesis of morphine and its congeners were in error. It was easy to write a new biosynthetic proposal starting with the benzylisoquinoline alkaloid **10**, then unknown, but later found to be a common natural product, reticuline. I proposed that **10** was oxidized in *Papaver somniferum* to give the dienone **11**, an exact analogue of the primary step of formation of Pummerer's ketone. By ring closure this would give **12**. Reduction of **12** to the allylic alcohol **13** and elimination of water would then furnish thebaine **14**. The steps from thebaine to codeinone **15a** then codeine **15b** on to morphine **16** were logical.

Later, in collaboration with Theodore (Ted) Cohen,[11] I

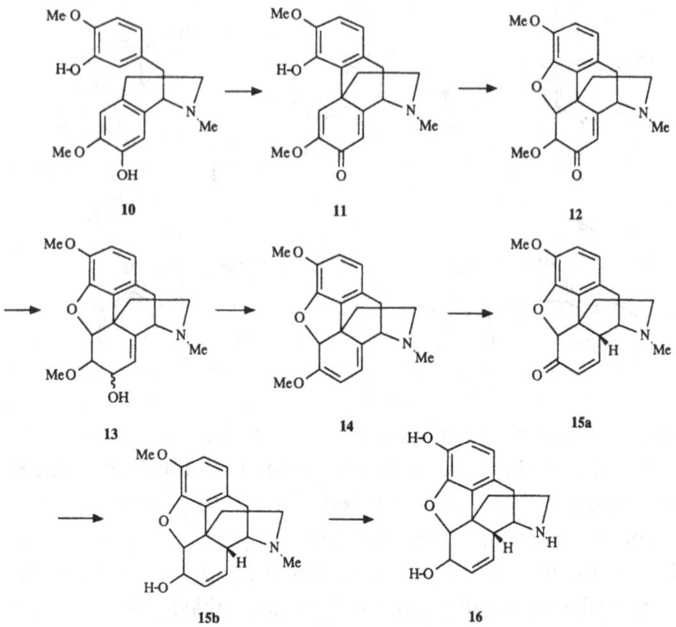

wrote an article on the biosynthesis of natural products by the coupling of phenolate radicals in pairs to make O–C or C–C bonds. However, we always followed the rules of *ortho-* or *para*-coupling. The theory was applied particularly to phenolic alkaloids. In order to explain what was formally *meta*-coupling, we postulated *para*-coupling followed by a dienone–phenol rearrangement. For certain alkaloids which contained only a benzene ring, we postulated a dienol–benzene rearrangement in biosynthesis before it had even been seen in the chemical laboratory. In fact, all of our proposals, including the existence of a new class of ketonic and phenolic alkaloids, proved, over the years, to be correct as outlined in more detail later.

In the laboratory and associated greenhouse, we gave particular attention to the biosynthesis of the morphine alkaloids.[12–14] At first we purchased *Papaver somniferum* seeds and grew splendid poppies, but they contained no morphine alkaloids. However, when we went through the official channels we were able to obtain seeds which provided poppies which looked the same, but in fact contained, when they ripened, genuine morphine alkaloids.

Precursors such as reticuline **10** were synthesized labelled with ^{14}C (O and N methyl groups) and with ^{3}H in the aromatic nuclei. Labelling could also be done in the two 2-carbon bridges. We also synthesized from thebaine the key alkaloid **11** for the first time. Unlike the situation with Pummerer's ketone **11** did not close to **12** spontaneously. Later on, alkaloid **11** was isolated from a Brazilian plant. From correspondence with Prof. R. A. Barnes, we realized that the two were probably identical, which was confirmed by an exchange of specimens. So, we used thereafter the name salutaridine, given by our Brazilian colleagues.

Labelled reticuline was readily incorporated into thebaine **14**, codeinone **15a**, codeine **15b**, and morphine **16** without scrambling of the labels. Salutaridine is not present in detectable quantities in *Papaver somniferum*. However, when appropriately labelled, it is well incorporated into morphine alkaloids. Salutaridine can be readily reduced to two stereoisomeric allylic alcohols **17**, both of which are converted by mild acid catalysis (**17**, see arrows) to give thebaine **14**. The alkaloids **17**, **14**, **15a**, and **15b** were all shown to be precursors of morphine. This was of interest, because the earlier theory of Robinson suggested that unmethylated alkaloids were first assembled and methylation was a terminal stage of biosynthesis.

It has not yet been proven that the coupling of phenolate radicals takes place in the biosynthesis of phenolic alkaloids. However, there is no evidence against, and much circumstantial support for, this proposal. The brilliant work of Zenk[15–17] has shown that the oxidative step in morphine biosynthesis is carried out by a single P-450 type iron porphyrin enzyme, which has been isolated pure and cloned from the appropriate gene. This is a remarkable accomplishment. Further work should clarify the biosynthesis of many other natural products where phenolate radical coupling has been proposed.

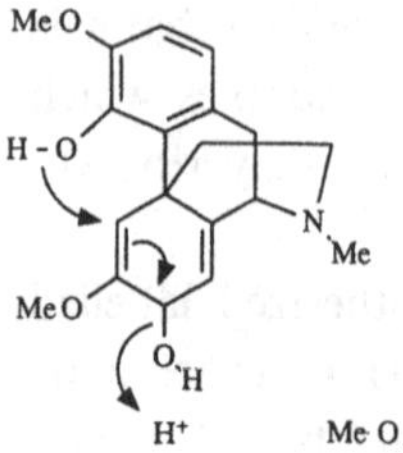

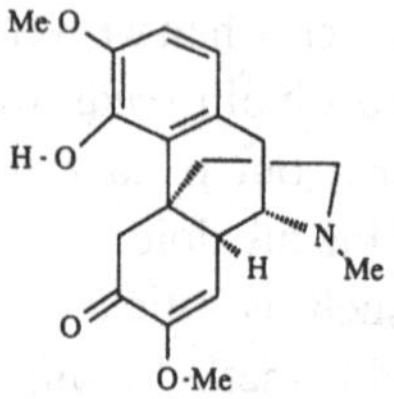

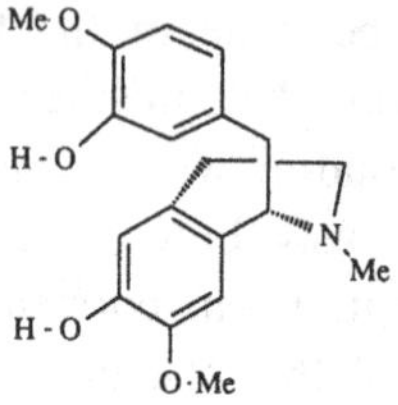

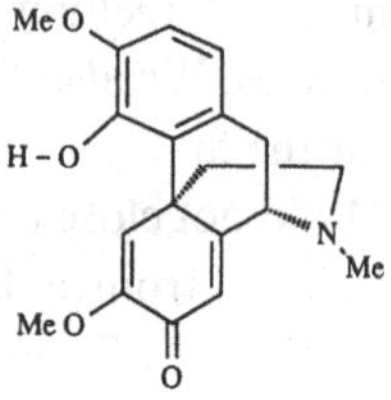

After the morphine studies, I turned my attention to sinomenine.[18,19] Robinson had proposed[20,21] that sinomenine **18** was formed by *meta*-coupling from 'protosinomenine' **19**. I considered that it was biosynthesized from reticuline via the enantiomer **20** of salutaridine **11**. In due course, labelled reticuline, 'protosinomenine' and other congeners were synthesized and fed to *Sinomenium acutum*. Only reticuline was incorporated. There was no significant incorporation of 'protosinomenine'. Evidently, the plant knew better than Sir Robert about *meta*-coupling. Whilst this work was in progress, Chu reported[22,23] the isolation of a new ketonic alkaloid **20** which was enantiomeric with salutaridine. I was delighted because, with a suitably labelled sample of this compound, its incorporation into sinomenine could be studied. Good incorporation was observed, and so 'protosinomenine', which had inspired much misleading work, could finally be laid to rest.

Mention has been made above of our proposals for phenolate

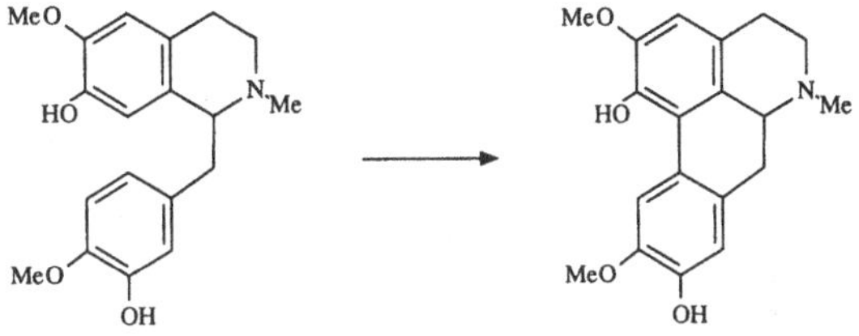

22 → **21**

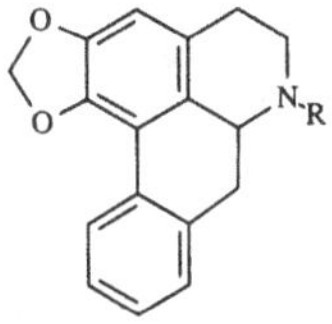

23 R = H
24 R = Me

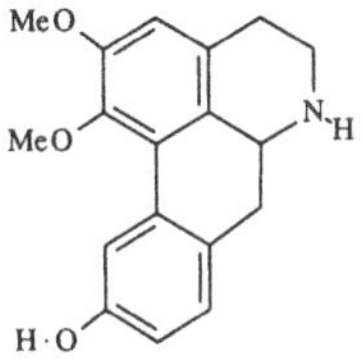

25

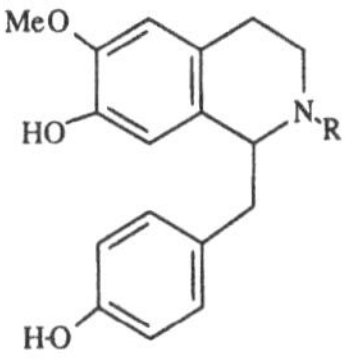

26 R = H
27 R = Me

coupling where the structures did not at first seem to fit the simple theory.[24–29] The biosynthesis of the alkaloids like **21** fits easily into the diphenolate radical coupling theory with **22** as the precursor. However, how could one explain the biosynthesis of alkaloids like anonaine **23**, roemerine **24**, and tudaranine **25**? We proposed that the precursor for these alkaloids was the diphenol **26** or **27**. By *ortho–para*-coupling, the dienones **28** and **29** would be formed. A dienone–phenol rearrangement would convert **28** into the carbon skeleton of **25**. The last step would be a simple *o*-methylation process to give **25** itself. Reduction of the ketone of **28** and **29** would give the dienols **30** and **31** respectively. We proposed that an acid-catalyzed dienol–benzene rearrangement would afford the carbon skeletons of **23** and **24**. The last step in the biosynthesis would be *ortho*-methoxyphenol cyclization to give the methylenedioxy group. This is a common biosynthetic process, although at the time there was no experimental evidence for it. Our later work,[30] as well as that of others,[31] provided firm support that this theory

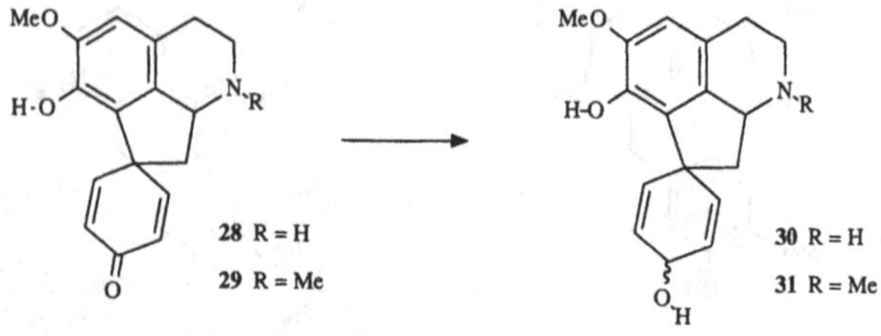

28 R = H
29 R = Me
30 R = H
31 R = Me

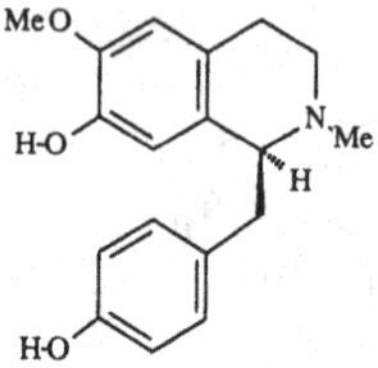

32
33 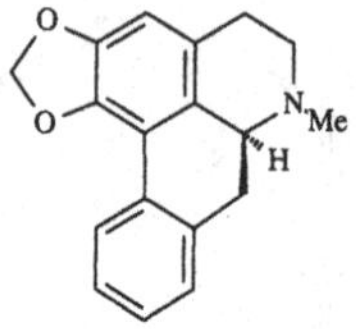

was correct. The best evidence came from doubly labelled precursors. In whole plant experiments, incorporations are often low. The constancy of the ratio of labels as well as the proof that the positions of the labels in the product are the same as in the precursor, provides valuable supplementary evidence of the biosynthetic process.

In later work, all these theoretical ideas were confirmed by

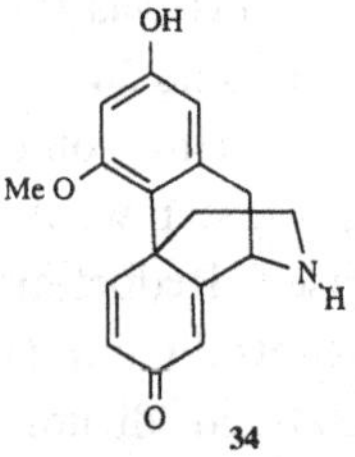

34
35 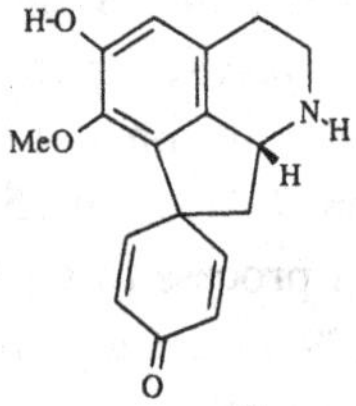
36 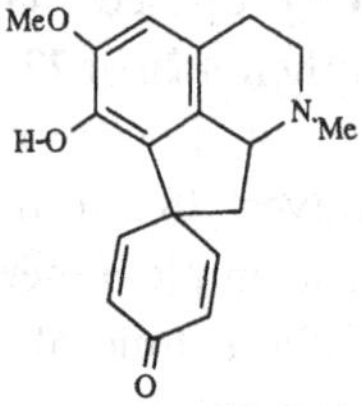
37 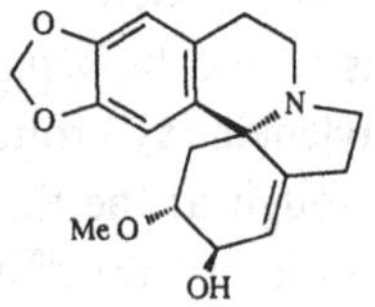
38 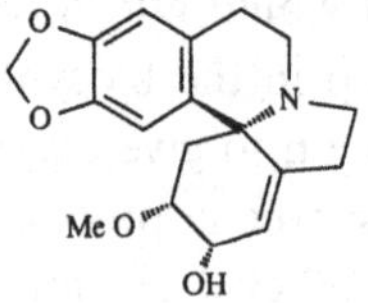
39

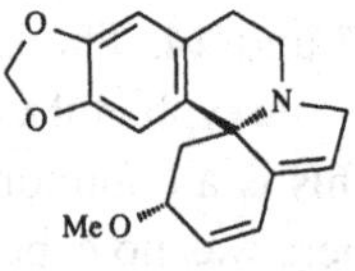

experiments on living plants. In particular we proved (+)-*N*-methylcoclaurine **32** was the precursor of (+)-roemerine **33**.

The interesting dienone alkaloid crotonosine was isolated and studied by Haynes and his collaborators.[32] It was provisionally formulated as **34**. This could have been in accord with phenolate coupling, but the resorcinol type ring seemed unusual. Further study showed[33–35] that crotonosine was really the dienone **35**. So what we had postulated to explain the formation of tudaranine **25** was shown, for the first time, to be true. Crotonosine was eventually correlated with tudaranine. Later many other dienone alkaloids were found in nature. An example from the same epoch as crotonosine is glaziovine **36**.[36] We carried out biosynthetic studies[37,38] on crotonosine and proved that the biosynthesis followed the phenolate coupling proposal with a demethylation–remethylation step (or its equivalent) as the last step in the biosynthesis.

My final studies on phenolate radical coupling were carried out on the *erythrina* alkaloids using small specimens of the beautiful flame tree: *Erythrina chrysta-galli*.[39–48] Typical *erythrina* alkaloids are erythratine **37**, epierythratine **38** and erysodine **39**.

The studies of the biosynthesis of *erythrina* alkaloids got off to a false start. In a beautiful experiment, A. I. Scott and colleagues showed[49,50] that the whole *erythrina* skeleton could be constructed in one simple phenolate coupling radical reaction.

Thus, oxidation of the symmetrical amine **40** gave a 35%

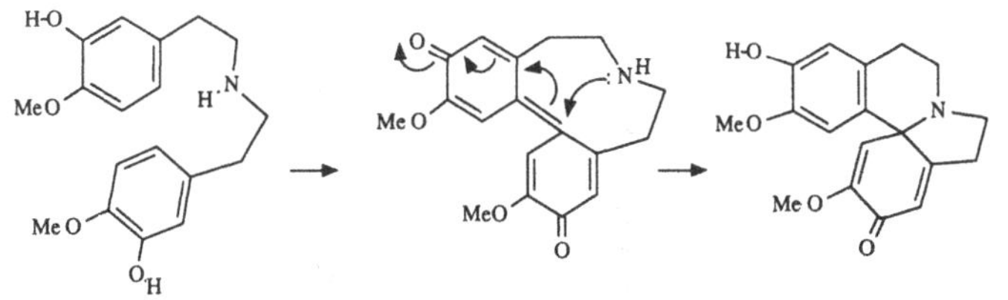

yield of the dienone **41** with **42** (*para–para* coupling) as the proposed intermediate. We showed that chromous chloride reduction of **41** gave the diphenol **43** (70% isolated) and that reoxidation with potassium ferricyanide gave back **41** (80% isolated). Since **43** would be the intermediate diphenol in the *para–para*-coupling process, this gave good support for the mechanism proposed. Finally, when we examined the alkaloids of *Erythrina chrysta-galli* we found another major alkaloid which had previously been overlooked. It was an expected derivative of **41** on the way to the formation of **37**, **38**, and **39**. In fact, it was the unsaturated ketone **44** which we called erythratinone. Naturally, erythratinone was well incorporated into the other *erythrina* alkaloids.

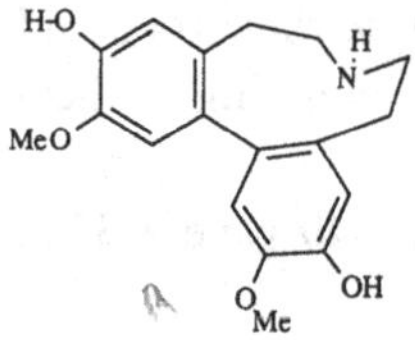

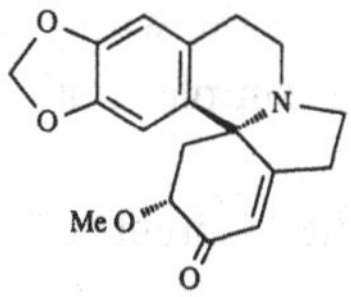

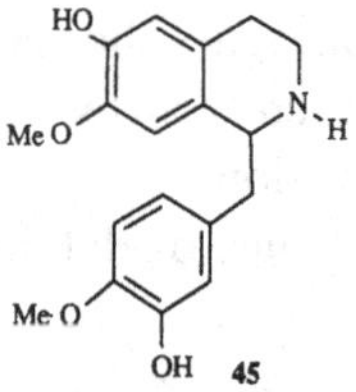

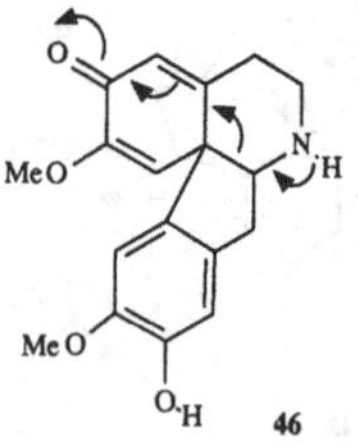

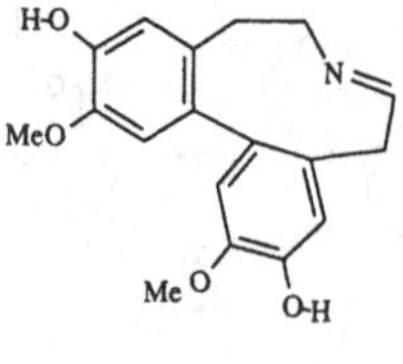

The diphenol **40** was synthesized in labelled form and fed to young specimens of *Erythrina chrysta-galli* several times. There was essentially no incorporation into *erythrina* alkaloids. Since we had never had a really negative answer before when we were following a real biosynthetic pathway involving phenolate radical coupling, we began to consider that perhaps the easy formation of **41** from **40** might be misleading. We supposed that the *erythrina* alkaloids might, in fact, be benzylisoquinoline derived. If we began with '*N*-norprotosinomenine' **45** then the same *para–para* coupling would give the dienone **46** whose fragmentation would afford the right carbon skeleton **47**.[46] This, on imine reduction, would give **43**; which (as already established) was readily converted to **41**. So, we synthesized labelled '*N*-norprotosinomenine' (and its congeners) and showed that it was an efficient precursor of *erythrina* alkaloids, whereas *N*-norreticuline and other congeners were not. So at last we found that the 'protosinomenine' (which was not) gave rise on *N*-demethylation to protoerythratinone.

I was pleased with this result because the plant had clearly indicated to us that the original, and chemically available, pathway was not correct. Further thought then gave the real pathway. I was also pleased that so much beautiful chemistry (by nature) came from reflections on the erroneous formula of Pummerer's ketone.

Exploration of the role of phenolic coupling in alkaloid biosynthesis was continued in a long series of elegant publications[51] by Dr D. S. Bhakuni who was associated with the work cited above on morphine biosynthesis. A. R. Battersby and his colleagues also made many important contributions in phenolic alkaloid biosynthesis and participated in the key studies of morphine biosynthesis.[52]

References

1. Pummerer, R., Puttfarcken, H. & Schopflocher, P. *Ber*. (1925), **58**, 1808.
2. Clemo, G. R., Haworth, R. D. & Walton, E. *J. Chem. Soc.* (1929), 2368.
3. Clemo, G. R., Haworth, R. D. & Walton, E. *J. Chem. Soc.* (1930), 2579.
4. Barton, D. H. R., Deflorin, A. M. & Edwards, O. E. *Chem. Ind.* (1955), 1039.
5. Barton, D. H. R. & Bruun, T. *J. Chem. Soc.* (1953), 605.
6. Foster, R. T., Robertson, A. & Healy, T. V. *J. Chem. Soc.* (1939), 1594.

7. Barton, D. H. R., Deflorin, A. M. & Edwards, O. E. *Chem. Ind.* (1955), 1039.
8. Barton, D. H. R., Deflorin, A. M. & Edwards, O. E. *J. Chem. Soc.* (1956), 530.
9. Pantilla, A., Fales, H. M. *J. Chem. Soc., Chem. Commun.* (1966), 656.
10. Taguchi, H., Sankawa, V. & Shibata, S. *Tetrahedron Lett.* (1966), 5211.
11. Barton, D. H. R. & Cohen, T. *Festschrift Arthur Stoll* (1957), 117.
12. Barton, D. H. R., Kirby, G. W., Steglich, W. & Thomas, G. M. *Proc. Chem. Soc.* (1963), 203.
13. Barton, D. H. R., Kirby, G. W., Steglich, W., Thomas, G. M., Battersby, A. R., Dobson, T. A. & Ramuz, H. *J. Chem. Soc.* (1965), 2423.
14. Barton, D. H. R., Bhakuni, D. S., James, R. & Kirby, G. W. *J. Chem. Soc., C* (1967), 128.
15. Zenk, M. H., Gerardy, R. & Stadler, R. *J. Chem. Soc., Chem. Commun.* (1989). 1725.
16. Loeffler, S., Stadler, R. & Zenk, M. H. *Tetrahedron Lett.* (1990), **31**, 4853.
17. Stadler, R. & Zenk, M. H. *Ann. Chem.* (1990), 555.
18. Barton, D. H. R., Kirby, A. J. & Kirby, G. W. *J. Chem. Soc., Chem. Commun.* (1965), 52.
19. Barton, D. H. R., Kirby, A. J. & Kirby, G. W. *J. Chem. Soc., C* (1968), 929.
20. Robinson, R. & Sugasawa, S. *J. Chem. Soc.* (1931), 3163.
21. Robinson, R. & Sugasawa, S. *J. Chem. Soc.* (1932), 789.
22. Chu, J.-H., Lo, S.-Y. & Chou, Y. L. *Acta Chim. Sinica.* (1964), **30**, 265.
23. Hsu, J.-S., Lo, S.-Y. & Chu, J.-H. *Scientia Sinica* (1964), **13**, 2016.
24. Barton, D. H. R. & Kirby, G. W. *Proc. Chem. Soc.* (1960), 392.
25. Barton, D. H. R. & Kirby, G. W. *J. Chem. Soc.* (1962), 806.
26. Barton, D. H. R., Kirby, G. W., Taylor, J. B. & Thomas, G. M. *Proc. Chem. Soc.* (1962), 179.
27. Barton, D. H. R., Kirby, G. W., Taylor, J. B. & Thomas, G. M. *J. Chem. Soc.* (1963), 4545.

28. Barton, D. H. R., Bhakuni, D. S., Chapman, G. M. & Kirby, G. W. *J. Chem. Soc., Chem. Commun.* (1966), 259.
29. Barton, D. H. R., Bhakuni, D. S., Chapman, G. M. & Kirby, G. W. *J. Chem. Soc., C* (1967), 2134.
30. Barton, D. H. R., Kirby, G. W. & Taylor, J. B. *Proc. Chem. Soc.* (1962), 340.
31. Battersby, A. R., Binks, R., Lawrie, W., Parry, G. V. & Webster, B. R. *J. Chem. Soc.* (1965), 7459.
32. Haynes, L. J. & Stuart, K. L. *J. Chem. Soc.* (1963), 1784 and 1789.
33. Haynes, L. J., Stuart, K. L., Barton, D. H. R. & Kirby, G. W. *Proc. Chem. Soc.* (1963), 280.
34. Haynes, L. J., Stuart, K. L., Barton, D. H. R. & Kirby, G. W. *Proc. Chem. Soc.* (1964), 261.
35. Barton, D. H. R., Haynes, L. J., Kirby, G. W. & Stuart, K. L. *J. Chem. Soc., C* (1966), 1676.
36. Cava, M. P., Nomura, K., Schlessinger, R. H., Buck, K. T., Douglas, B., Raffauf, R. T. & Weisbach, J. A. *Chem. Ind.* (1964), 282.
37. Haynes, L. J., Stuart, K. L., Barton, D. H. R., Bhakuni, D. S. & Kirby, G. W. *J. Chem. Soc., Chem. Commun.* (1965), 141.
38. Barton, D. H. R., Bhakuni, D. S., Chapman, G. M., Kirby, G. W., Haynes, L. J. & Stuart, K. L. *J. Chem. Soc., C* (1967), 1295.
39. Barton, D. H. R., James, R., Kirby, G. W., Turner, D. W. & Widdowson, D. A. *J. Chem. Soc., Chem. Commun.* (1966), 295.
40. Barton, D. H. R., James, R., Kirby, G. W. & Widdowson, D. A. *J. Chem. Soc., Chem. Commun.* (1967), 266.
41. Barton, D. H. R., James, R., Kirby, G. W., Turner, D. W. & Widdowson, D. A. *J. Chem. Soc., C* (1968), 1529.
42. Barton, D. H. R., Boar, R. B. & Widdowson, D. A. *J. Chem. Soc., C* (1970), 1208.
43. Barton, D. H. R., Boar, R. B. & Widdowson, D. A. *J. Chem. Soc., C* (1970), 1213.
44. Barton, D. H. R., Jenkins, P. N., Letcher, R., Widdowson, D. A., Hough, E. & Rogers, D. *J. Chem. Soc., Chem. Commun.* (1970), 392.
45. Barton, D. H. R., Gunatilaka, A. L., Letcher, R. M., Lobo, A. M. F. T. & Widdowson, D. A. *J. Chem. Soc., Perkin Trans. 1*

(1973), 875.
46. Barton, D. H. R., Potter, C. J. & Widdowson, D. A. *J. Chem. Soc., Perkin Trans. 1* (1974), 346.
47. Barton, D. H. R., Bracho, R. D., Potter, C. J. & Widdowson, D. A. *J. Chem. Soc., Perkin Trans. 1* (1974), 861.
48. Barrett, A. G. M., Barton, D. H. R., Franckowiak, G., Papaioannou, D. & Widdowson, D. A. *J. Chem. Soc., Perkin Trans. 1* (1979), 662.
49. Gervay, J. E., McCapra, F., Money, T., Sharma, G. M. & Scott, A. I. *J. Chem. Soc., Chem. Commun.* (1966), 142.
50. Mondon, A. & Ehrhardt, M. *Tetrahedron Lett.* (1966), 2557.
51. a) Bhakuni, D. S. & Jain, S. *Tetrahedron* (1981), **37**, 3175. b) Bhakuni, D. S. & Jain, S. *Tetrahedron* (1981), **37**, 3171. c) Bhakuni, D. S. & Jain, S. *J. Chem. Soc., Perkin Trans. 1* (1981), 2598. d) Bhakuni, D. S., Jain, S. & Singh, A. N. *Phytochem.* (1980), **19**, 2347. e) Bhakuni, D. S., Jain, S. & Singh, R. S. *Tetrahedron* (1980), **36**, 2525. f) Bhakuni, D. S., Singh, A. N. & Jain, S. *J. Chem. Soc., Perkin Trans. 1* (1978), 1318. g) Bhakuni, D. S. & Singh, A. N. *J. Chem. Soc., Perkin Trans. 1* (1978), 618. h) Bhakuni, D. S., Jain, S. & Singh, A. N. *J. Chem. Soc., Perkin Trans. 1* (1978), 380. i) Bhakuni, D. S., Tewari, S. & Kapil, R. S. *J. Chem. Soc., Perkin Trans. 1* (1977), 706 and references cited therein.
52. a) Battersby, A. R. *Ciba Found. Symp.* (1978), **53**, 25. b) Battersby, A. R., Jones, R. C. F. & Kazlauskas, R. *Tetrahedron Lett.* (1975), **22**, 1873. c) Battersby, A. R., Ramage, R., Cameron, A. F., Hannaway, C. & Šantavy, F. *J. Chem. Soc. C* (1971), 3514. d) Battersby, A. R., Foulkes, D. M., Hirst, M., Parry, G. V. & Staunton, J. *J. Chem. Soc. C* (1968), 510. e) Battersby, A. R., Martin, J. A. & Brochman-Hanssen, E. *J. Chem. Soc. C* (1967), 1785. f) Battersby, A. R. *J. Chem. Soc., Chem. Commun.* (1967), 483. g) Battersby, A. R. *J. Chem. Soc.* (1965), 3323.

3

Nitrite photolysis (the Barton reaction)

The last important steroidal hormone of the adrenal cortex to be characterized was aldosterone **1a**. This substance was only available in minute amounts from natural sources and the determination of structure by Reichstein and his colleagues[1,2] was a masterpiece of collaboration between University (Basel) and Industry (Ciba-Geigy). The masked aldehyde group at C_{18} is an unusual feature for a steroidal molecule and at once posed interesting problems of synthesis. There are very few natural steroids which are substituted at C_{18}, so a partial synthesis did not, at first, seem very practical.

Aldosterone controls the electrolyte balance of mammalian systems, and so there was tremendous interest in obtaining enough aldosterone to examine in detail its biological proper-

3 R = H
4 R = NO

5 X = N-OH
6 X = O

1a R = R′ = H
1b R = H, R′ = Ac
2a R = OH, R′ = H
2b R = OH, R′ = Ac

ties. In the late 1950s and in the 1960s an enormous effort was devoted to the total, and to the partial, synthesis of aldosterone.

I was interested in the problem because of my association with the Research Institute for Medicine and Chemistry (RIMAC) in Cambridge, MA. This institute was founded and funded by the Schering Corporation (later the Schering-Plough Corp.). The Director was Dr M. M. Pechet, and I was chosen to lead a group devoted to the chemical synthesis of biologically important compounds. I agreed to this on the condition that we should always do original chemistry on the way to our synthetic goals.

We chose aldosterone **1a** as our first synthetic objective, with 17-α-hydroxyaldosterone **2a** as a second, and more difficult target. We decided to start with a readily available corticosteroid derivative such as **3** (corticosterone acetate). Such compounds were available from the degradation of cholic acid or of diosgenin. But we needed to introduce the aldehyde function at C_{18}. To do this we needed to invent a new reaction.[3,4] Little was known at that time about selective attack on unactivated methyl groups. It was clear to me that, if we could convert the axial 11β-hydroxyl group into an alkoxy radical, this would abstract hydrogen from one of the axial methyl groups at C_{18} and C_{19}. Alkoxy radicals usually stabilize themselves by hydrogen abstraction. If the attack occurred mainly towards C_{18}, to make a primary radical, we would need to capture this radical in such a way as to give an aldehyde or the precursor of an aldehyde. The solution to the problem came easily because my mind was prepared.

My mind was prepared from my knowledge of the pyrolysis of chlorinated hydrocarbons (Chapter 1) and related subjects. The work of the late Professor E. W. R. Steacie, who eventually became Director of the National Research Council of Canada, had showed that the pyrolysis of alkyl nitrites in the gas phase gave NO and an alkoxy radical in a unimolecular reaction. The temperatures required for this reaction were much too high for use in steroidal chemistry. There was little work on the photolysis of nitrites and some dispute on the products formed. I decided that the photolysis of the 11β-nitrite **4** of corticosterone acetate would give the required alkoxy radical which, if it attacked C_{18}, would furnish the C_{18} radical which should be captured by the NO liberated in the first step. It was well known that carbon radicals are efficiently captured by NO from the physico-chemical studies of Hinshelwood and Steacie during the 1930s. Thus a nitroso group would be introduced at C_{18}. Such nitroso groups with α-H readily isomerize to the corresponding oximes **5**. Thus there was a simple conceptual route, carried out at room temperature using 350 nm irradiation of an easily prepared nitrite, to furnish the equivalent of an aldehyde. The conversion of relatively unhindered oximes to the corresponding carbonyl compounds is easily carried out by treatment of the oxime with nitrous acid in acetic acid containing some water. So this reaction, applied to **5**, should afford the desired aldehyde **6**. Scheme 1 summarizes the proposed mechanism for the general case.

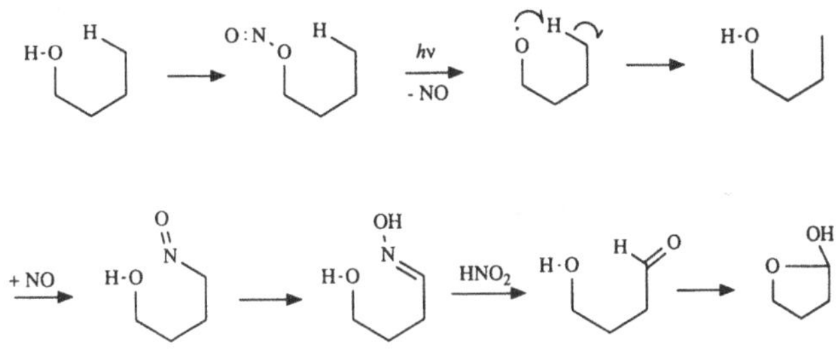

Scheme 1

In reality, it required only a few weeks to convert **3** into aldosterone acetate **1b**.[5–7] The acetate is a biological equivalent to aldosterone and easier to handle and store. The overall yield of crystalline acetate was about 15%. However, this was satisfactory for the preparation of 60 g of aldosterone acetate

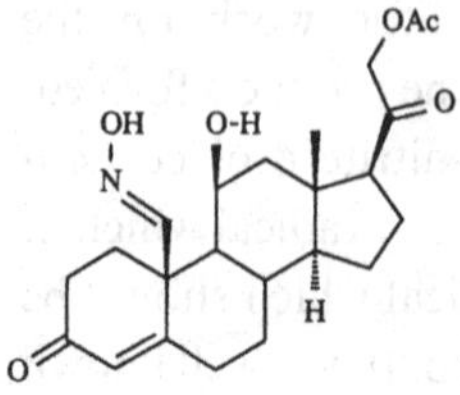

7

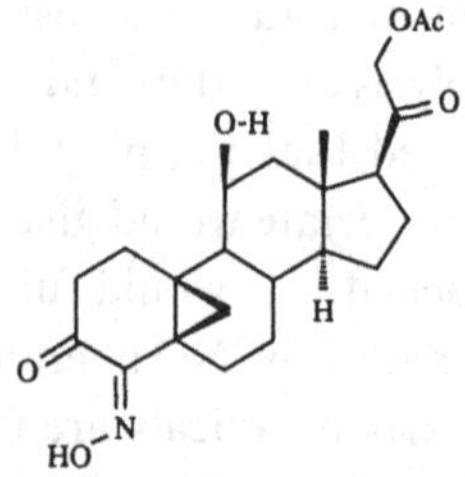

8

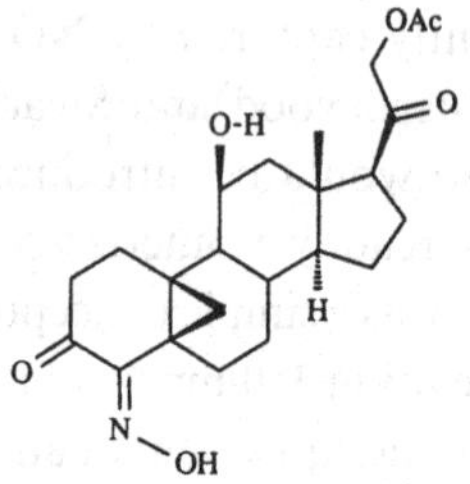

9

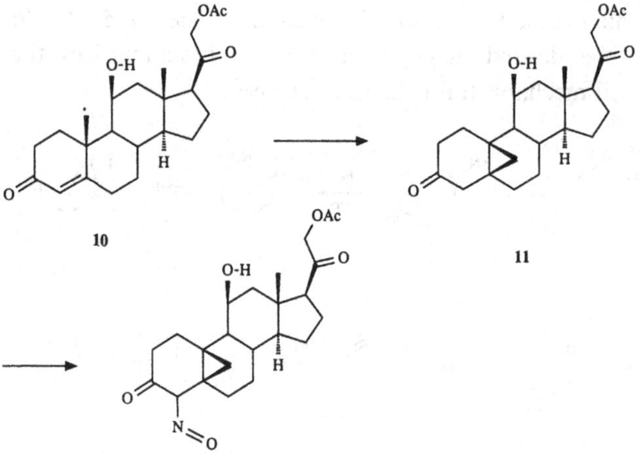

10 11

12

1b at a time when the world's supply was only a few milligrams.

The relatively poor overall yield was due principally to the partitioning of the 11β-alkoxy radical between the 18 and 19 methyl groups. Neglecting conformation effects due to the 4(5)-

double bond and the *trans*-C-D hydrindane fusion, the 11β-alkoxy radical is equidistant to the two methyls. In fact, the radical attack was about equal on C_{18} and C_{19}. However, attack on C_{19} did not give the expected oxime **7**. Instead, a mixture of the oximino-ketones **8** and **9** was formed. From the structures of **8** and **9**, it was clear that the initially formed carbon radical **10** had cyclized to radical **11** which then reacted with NO to give nitroso-derivative **12** from which, by isomerization, **8** and **9** were formed.[8] The cyclization of radical **10** to **11** is surprising since it is the reverse reaction which is normally seen. We attributed this to the stabilization of the radical by conjugation of the radical center with the ketone at C_3. It was of interest to prepare the C_{19} isomer **13** of aldosterone acetate **1b** and also 19-noraldosterone acetate **14**. The problem of the cyclization of radical **10** was resolved by starting with the easily prepared (from **3**) diketal **15**. Conversion of **15** to its nitrite and photolysis gave isomers at C_{18} and C_{19} which could be

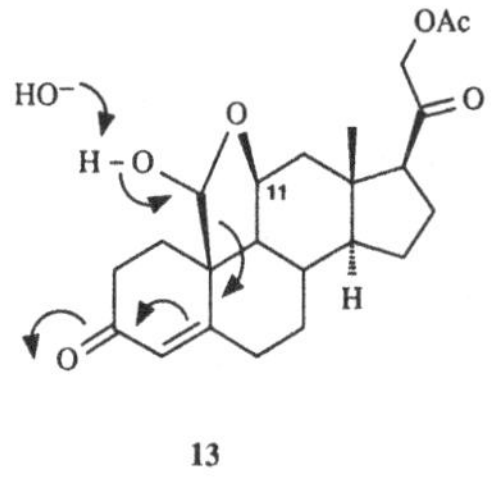

13

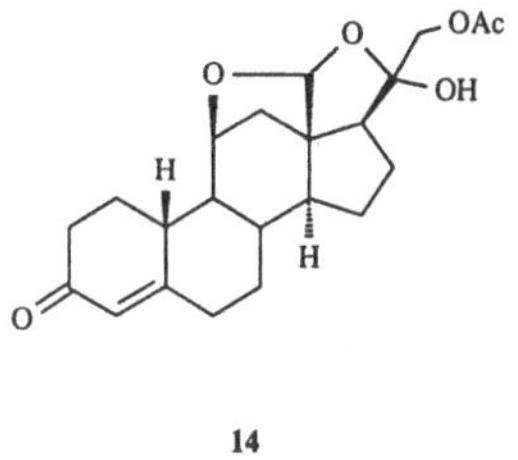

14

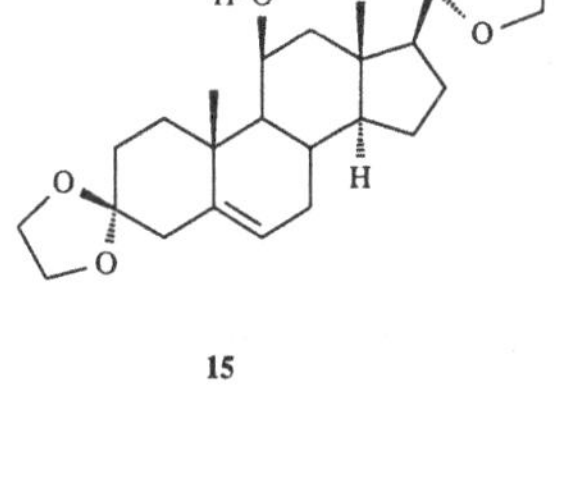

15

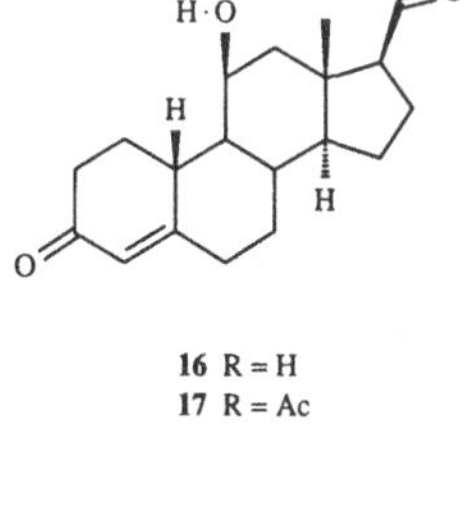

16 R = H
17 R = Ac

separated. Treatment with nitrous acid and mild acid-catalyzed hydrolysis of the ketal functions gave, after reacetylation, a further synthesis of aldosterone acetate **1b**. From the other oxime, the desired isomer of aldosterone acetate, **13** was obtained.[9] However, it had no significant biological activity. From the structure of **13** it was possible to predict that mild treatment with alkali should afford the 19-norsteroid **16** (**13**: see arrows), easily reacetylated to **17**. When **17** was subject to the same sequence used to make aldosterone acetate **1b**, it readily gave the desired 19-noraldosterone acetate **14**. Other 19-nor syntheses were also developed.[10,11]

The importance of conformation in selective attack of the 11β-alkoxy radical on C_{18} and C_{19} was nicely illustrated by the photolysis of the nitrite **18** derived from the dienone **19**. This afforded only C_{18} attack.[7] Treatment of the derived oxime with nitrous acid gave 1-dehydro-aldosterone acetate **20**. This is a convenient precursor for labelled aldosterone acetate since selective hydrogenation with tritium gives 1,2-ditritiated aldosterone acetate.

In order to synthesize 17-α-hydroxyaldosterone **2a**,[12] a side chain protection was needed. Otherwise the 18-carbon radical tended to furnish a methylene group and open the 13–17 bond in ring D. A convenient protection for the corticoid side chain is the *bis*-ketal derived from formaldehyde. Thus *bis*methylenedioxy-1-dehydrohydrocortisone **21** was converted to its nitrite and photolyzed in the usual way. The only product (60 % isolated yield) was the desired isomer **22**. When this was treated with nitrous acid, an unusual sequence of reactions took place. The normal nitrosation intermediate underwent ring closure to **24** followed by oxonium ion formation **25** as indicated. The oxonium ion **25** then opened one of the adjacent methylenedioxy groups (**25** → **26**) which on hydration with water afforded the isolated product **27**. Acetylation of **27** with acetic anhydride and acid catalysis gave the triacetate **28** in ≈ 80 % overall yield from the oxime **22**. Mild alkaline hydrolysis

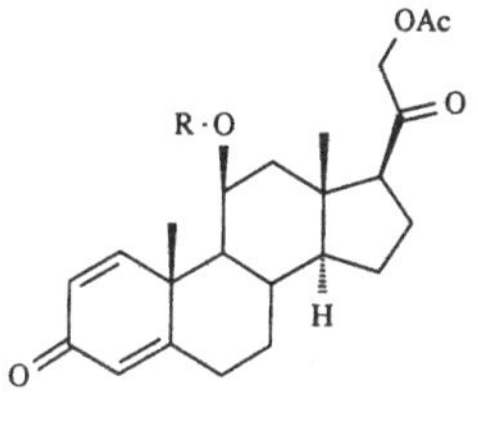

18 R = NO
19 R = H

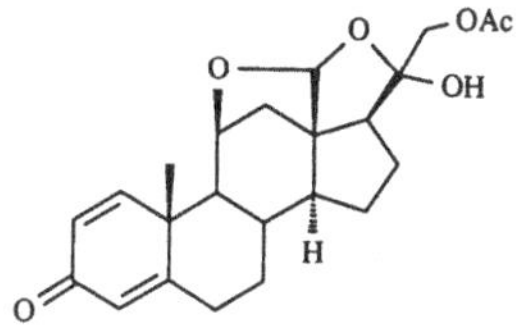

20

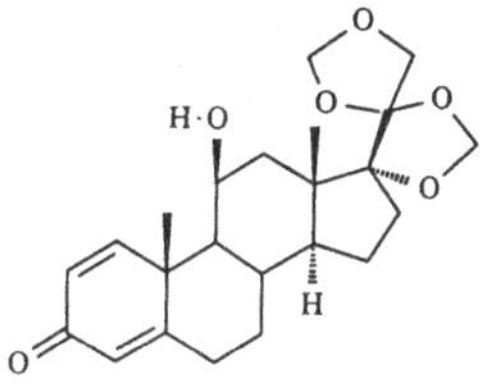

21

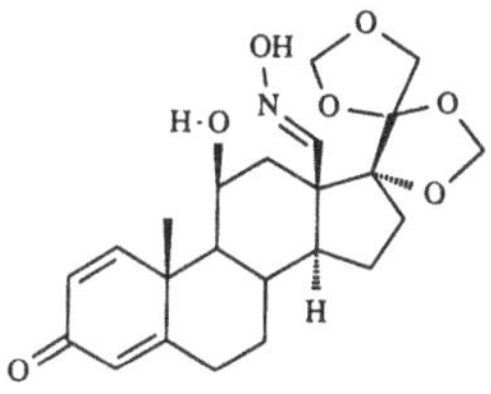

22

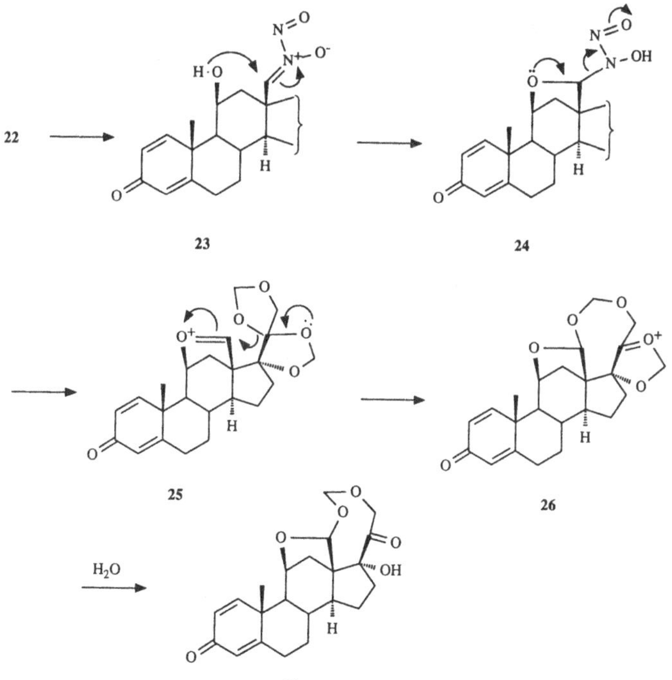

22 → **23** → **24**

→ **25** → **26**

H_2O → **27**

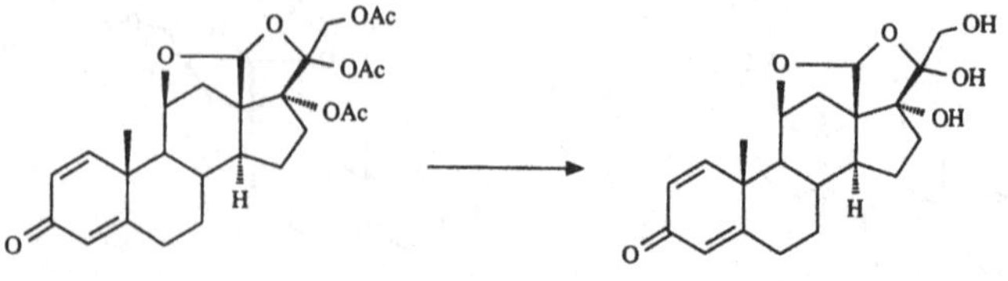

28

29

of **28** then gave the desired 1-dehydro-17α-hydroxyaldosterone **29**. Selective acetylation and hydrogenation afforded the synthetic objective, 17-α-hydroxyaldosterone acetate **2b**, for the first time.

An improved synthesis[7] of aldosterone acetate **1b** was based on the readily available dienone **30**. Conversion to the nitrite and photolysis affords the oxime **31** with no attack on C_{19}. On warming in iso-propyl alcohol this oxime **31**, cyclized smoothly to the nitrone **32** with the loss of water. This nitrone is at the right oxidation level to rearrange to the 21-acetoxyimine. The overall yield of nitrone from **30** was 55 %.

Acetylation of the nitrone **32** gave the enol acetate **33** which

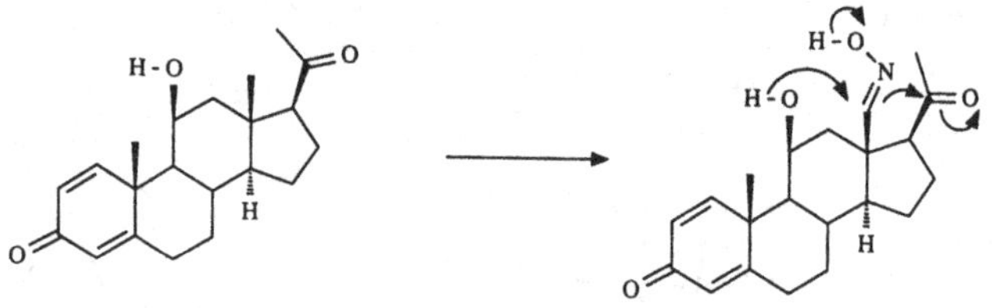

30

31

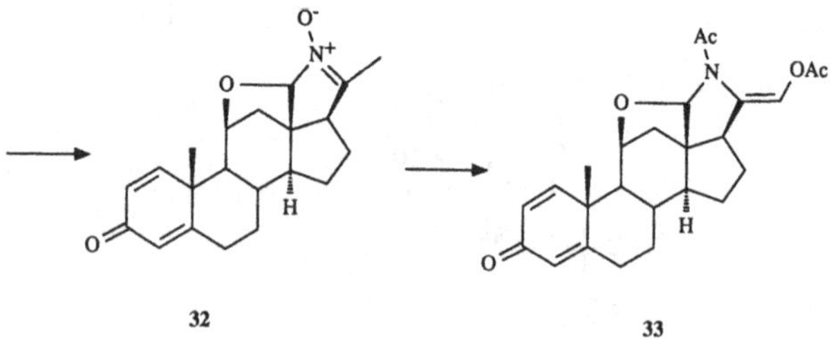

32

33

on hydration with aqueous acetic acid afforded **34** in high yield. Nitrosation with N_2O_4–NaOAc in chloroform gave the nitroso derivative **35** which on heating in dioxane afforded the acetate **36**. Hydrolysis with aqueous acetic acid then gave the desired 1-dehydroaldosterone acetate **20**.

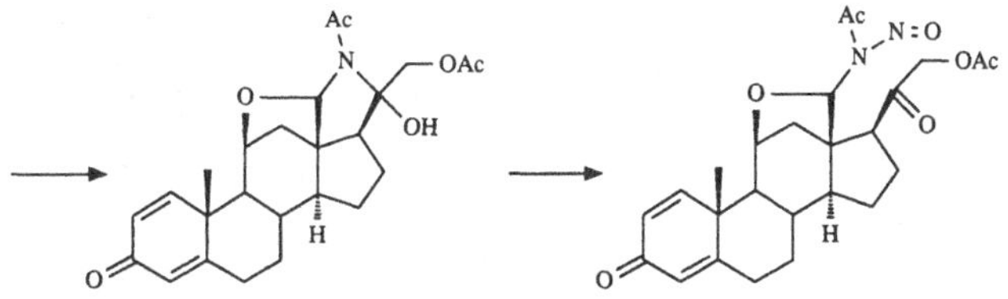

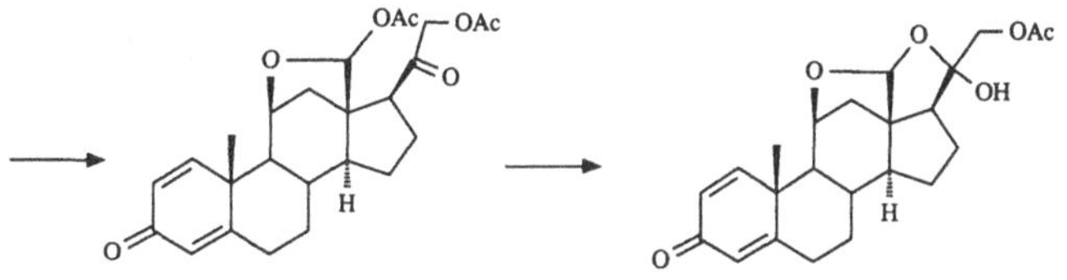

The report that 18-hydroxycorticosterone acetate **37** and its 11-deoxy derivative **38** were biologically important stimulated considerable synthetic effort. In an earlier investigation[13] we had accidentally discovered that nitrite photolysis in the presence of oxygen could afford a good yield of the nitrate

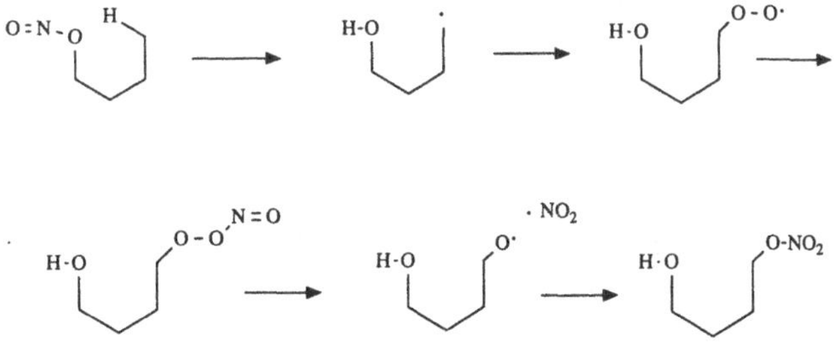

Scheme 2

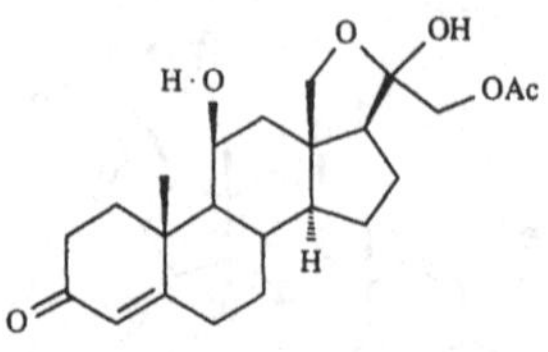

37 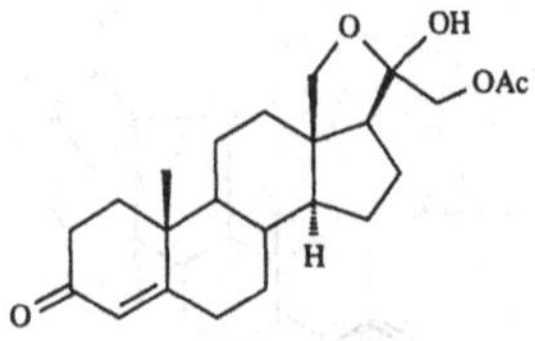 **38**

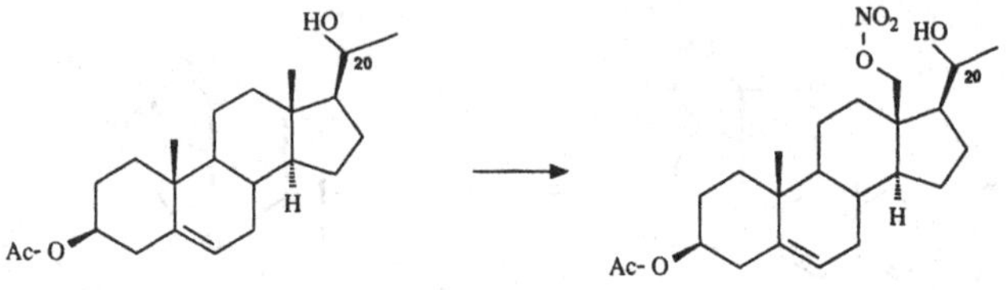

39 (20 α or β) **40** (20 α or β)

corresponding to the normal oxime. This reaction can be represented as in Scheme 2. The nitrate, although much more hydrolytically stable than a nitrite, is readily reduced to the corresponding alcohol. In fact, the nitrate function is a good protecting group for an alcohol.

For the synthesis[14] of 11-deoxy-corticosterone acetate, a convenient starting material was the readily available alcohol **39**. On photolysis under oxygen, the corresponding nitrite gave the desired nitrate **40**. Standard oxidation, hydrolysis and Oppenauer oxidation gave 18-nitrate **41** which was readily reduced under mild conditions with zinc dust to give the 18-

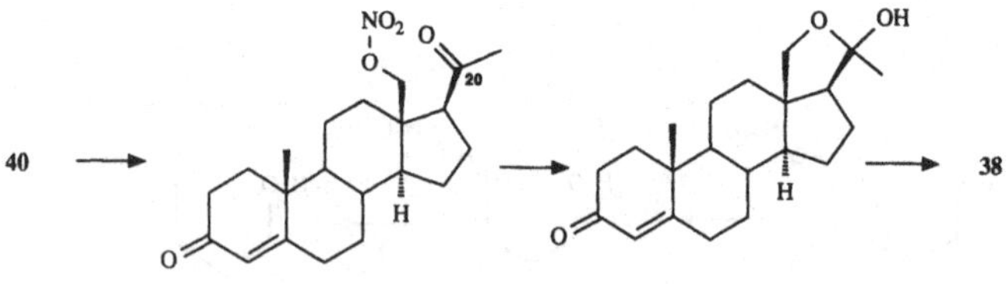

40 → **41** → **42** → **38**

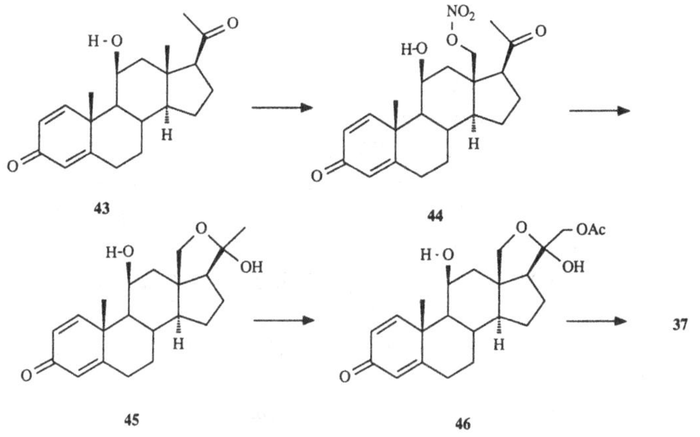

hydroxy derivative **42**. Acetoxylation of **42** with lead tetra-acetate then afforded the target molecule **38**.

For 18-hydroxycorticosterone acetate[14] the starting material was the readily available **43**. Conversion to nitrite and photolysis under oxygen gave the desired nitrate **44** in satisfactory yield. Reduction gave the 18-hydroxy function **45** which on acetoxylation as above gave **46**. Selective hydrogenation with Wilkinson's catalyst then gave the desired molecule **37**.

A detailed study of the mechanism of nitrite photolysis[15] confirmed the ideas used to conceive the existence of the reaction. The carbon radical generated by the photolysis is free because it can be captured by radicophylic reagents permitting other substituents[16,17] to be introduced at C_{18}.

It was easy to carry out at the right time (1960–1965) interesting and original chemistry by nitrite photolysis. An agreeable relationship with the Schering Corp. produced a number of elegant publications[18–24] of which I am proud.

It will be clear that nitrite photolysis is compatible with many of the sensitive functions found in complex steroids. It provides a useful means of molecular manipulation using radical chemistry.

References

1. Simpson, S. A., Tait, J. F., Wettstein, A., Neher, R., von Euw, J., Schindler, O. & Reichstein, T. *Experientia* (1954), **10**, 132.
2. Simpson, S. A., Tait, J. F., Wettstein, A., Neher, R., von Euw, J., Schindler, O. & Reichstein, T. *Helv. Chim. Acta* (1954), **37**, 1200.
3. Barton, D. H. R., Beaton, J. M., Geller, L. E. & Pechet, M. M. *J. Am. Chem. Soc.* (1960), **82**, 2640.
4. Barton, D. H. R., Beaton, J. M., Geller, L. E. & Pechet, M. M. *J. Am. Chem. Soc.* (1961), **83**, 4076.
5. Barton, D. H. R. & Beaton, J. M. *J. Am. Chem. Soc.* (1960), **82**, 2641.
6. Barton, D. H. R. & Beaton, J. M. *J. Am. Chem. Soc.* (1961), **83**, 4083.
7. Barton, D. H. R., Basu, N. K., Day, M. J., Hesse, R. H., Pechet, M. M. & Starratt, A. N. *J. Chem. Soc., Perkin Trans. 1* (1975), 2243.
8. Barton, D. H. R. & Beaton, J. M. *J. Am. Chem. Soc.* (1961), **83**, 4083.
9. Barton, D. H. R. & Beaton, J. M. *J. Am. Chem. Soc.* (1962), **84**, 199.
10. Barton, D. H. R. & Akhtar, M. *J. Am. Chem. Soc.* (1962), **84**, 1406.
11. Barton, D. H. R. & Akhtar, M. *J. Am. Chem. Soc.* (1964), **86**, 1528.
12. Akhtar, M., Barton, D. H. R., Beaton, J. M. & Hortmann, A. G. *J. Am. Chem. Soc.* (1963), **83**, 1512.
13. Allen, J., Boar, R. B., McGhie, J. F. & Barton, D. H. R. *J. Chem. Soc., Perkin Trans. 1* (1973), 2402.
14. a) Barton, D. H. R., Day, M. J., Hesse, R. H. & Pechet, M. M. *J. Chem. Soc., Perkin Trans. 1* (1975), 2243. b) Barton, D. H. R., Basu, N. K., Day, M. J., Hesse, R. H., Pechet, M. M. & Starratt, A. N. *J. Chem. Soc., Perkin Trans. 1* (1975), 2252.
15. Barton, D. H. R., Hesse, R. H., Pechet, M. M. & Smith, L. S. *J. Chem. Soc., Perkin Trans. 1* (1979), 1159.
16. Akhtar, M., Barton, D. H. R. & Sammes, P. G. *J. Am. Chem. Soc.* (1964), **86**, 3394.
17. Akhtar, M., Barton, D. H. R. & Sammes, P. G. *J. Am. Chem.*

Soc. (1965), **87**, 4601.
18. Nussbaum, A. L., Carlton, F. E., Oliveto, E., Townley, E., Kabaskalian, P. & Barton, D. H. R. *J. Am. Chem. Soc.* (1960), **82**, 2973.
19. Reimann, H., Capomaggi, A. S., Strauss, T., Oliveto, E. P. & Barton, D. H. R. *J. Am. Chem. Soc.* (1961), **83**, 4481.
20. Nussbaum, A. L., Carlton, F. E., Oliveto, E., Townley, E., Kabaskalian, P. & Barton, D. H. R. *Tetrahedron* (1962), **18**, 373.
21. Robinson, C. H., Gnoj, O., Mitchell, A., Wayne, E., Townley, E., Kabaskalian, P., Oliveto, E. P. & Barton, D. H. R. *J. Am. Chem. Soc.* (1961), **83**, 1771.
22. Nussbaum, A. L., Robinson, C. H., Oliveto, E. P., Barton, D. H. R. *J. Am. Chem. Soc.* (1961), **83**, 2400.
23. Robinson, C. H., Mitchell, A., Oliveto, E. P., Beaton, J. M. & Barton, D. H. R. *J. Org. Chem.* (1962), **27**, 20.
24. Robinson, C. H., Gnoj, O., Mitchell, A., Oliveto, E. P. & Barton, D. H. R. *Tetrahedron* (1965), **21**, 743.

4

Radical deoxygenation (the reaction of Barton and McCombie)

The Barton–McCombie radical deoxygenation reaction was invented because of prior experience that I had had with the chemistry of thionobenzoates. In 1970, the late Professor Achmatowicz, one of Poland's greatest chemists, asked me to accept his elder son, Selim Achmatowicz, as a post-doctoral fellow at Imperial College. I at once accepted and Selim arrived with his own fellowship. What I did not know was that Selim was a wonderful bridge player. So good was he that he played in the Polish International Bridge Team on a regular basis. To keep at the level required, you obviously had to play several hours of bridge per day.

I asked him to study methods for the preparation of cholesterol thionobenzoate **1**. Some 20 years before I had been interested in the rearrangement of 5α,6β- into 5β,6α-dibromocholestanol esters.[1] This reaction had been in the literature for a long time, but the stereochemistry of the two dibromides had not been determined and the mechanism of the reaction had not been investigated. This rearrangement is best regarded as having a bridged bromonium ion in intimate pair with bromide anion. It was studied as an early example of conformational analysis. Later we extended[2] these reversible α,β-rearrangements to other functional groups like bromohydrin esters. We wanted a method to make thionobenzoates to study the participation of this functional group in irreversible α,β-, α,γ-

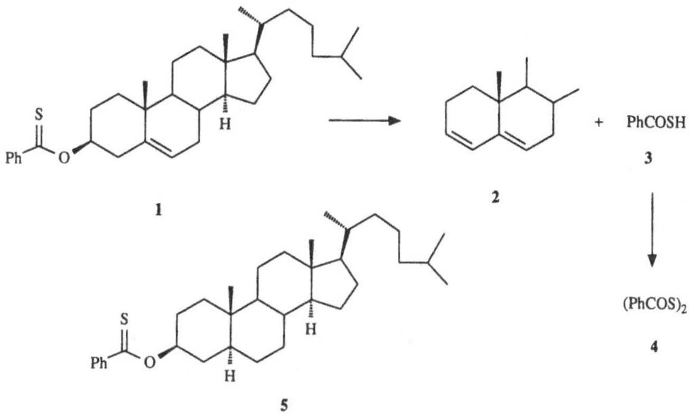

and eventually α,ω-rearrangements. This study was finally accomplished.[2]

Selim Achmatowicz did not have many hours to spend in practical work each day. After 2–3 hours in the morning, he usually disappeared for the rest of the day to practise bridge. I could not protest. He had his own money and a good excuse. Finally, one Thursday we were able to prepare a beautiful yellow solution of **1** in ether. However, Selim had to leave at once for a major bridge engagement in some far distant land. He promised to be back the following Tuesday. It was my habit to visit the laboratory two times each day at the hours of tea-time. As I passed by the flask on that Friday afternoon, I thought that some of the color had faded. A visit on Saturday confirmed this impression. By Monday afternoon, the solution was colorless. On the return of our bridge hero, I told him that he probably had a new photochemical reaction on his hands. The products formed quantitatively were cholesta-3,5-diene **2** and thiobenzoic acid **3**. The latter was rapidly oxidized by air to the disulfide **4**. We quickly established that cholestanol thionobenzoate **5** is stable to visible light under the same reaction conditions. This, and many other facts, illustrated the limitations of the reaction. The mechanism was studied in col-

H H O Ph S· 6 → O Ph SH 7 → 2 + 3

laboration with Sir George (now Lord) Porter and Dr J. Wirz.[3] It was confirmed that a high quantum yield was obtained and that the elimination was like a Norrish type II fragmentation involving the lowest $n \rightarrow \pi^*$ triplet. This means that the 4β-hydrogen in **1** is abstracted by radicaloid sulfur **6** to give a 4β-radical **7** which fragments to the observed products. Many further examples of this reaction were discovered.[4-6] In all cases, the hydrogen abstracted by the radicaloid sulfur was allylic, benzylic or α- to a heteroatom such as oxygen or nitrogen. For example, styrenes **9** were formed in very high yield by the irradiation of the corresponding thionobenzoyl derivatives **8** of the appropriate β-phenylethanol with a tungsten lamp at room temperature.

It is amusing that, without the game of bridge, this new reaction would still be waiting to be discovered, for we would have evaporated the ether on the same day as the preparation, or at least the next day! We called it the 'game-of-bridge reaction'.

To the best of my knowledge, this reaction has never been used in synthesis. This is a pity because it operates under very

X O S Ph 8 → X 9

mild conditions and in very high yield where the C–H bond which is broken is weak enough.

During the period 1954 to 1977, I was a regular consultant for the Schering-Plough Corporation. This agreeable relationship was renewed in 1986 when I took up my present appointment at Texas A&M and continues to this day. When it became clear that the family of aminoglycoside antibiotics, and especially the gentamycins, was going to be very important in the treatment of bacterial disease, I started to become interested in carbohydrate chemistry. Carbohydrate chemists like to cultivate the mystique that their reactions are esoteric and not attainable without many years of training. This, of course, is nonsense. I found it no more difficult to work with carbohydrates than to manipulate the sesqui, di- and triterpenoids or steroids. It was all a question of protecting groups.

During the course of the Schering-Plough work on the gentamycins, it became important to remove certain secondary hydroxyl groups. In some cases, the biological activity increased because the bacteria could no longer detoxify the antibiotics by acylation or phosphorylation. However, it is not always possible to remove efficiently the hydroxyl group by ionic reactions. Neighbouring group participation, or steric hindrance, may make ionic methods of deoxygenation impossible. I decided to invent a new reaction for this purpose.

It seemed to me that a deoxygenation reaction producing the corresponding carbon radical would have many advantages. Although neighbouring group participation in radical reactions is known, it is not seen with the usual protecting groups employed in carbohydrate chemistry. Also, radical reactions are much less liable to steric hindrance because, unlike ionic species, carbon radicals are not solvated. So the problems of ionic chemistry would be resolved. However, the carbon–oxygen bond in a secondary alcohol is a strong bond and a new reaction would be needed to break it to give a carbon radical.

I, at once, thought of the game-of-bridge reaction. There is no corresponding reaction of benzoates. Therefore, there is a driving force in the thionobenzoate to thiobenzoate transformation (C=S → C=O) that is enough to cause the elimination to give the second double bond. In the absence of photochemical activation, what device should we use? I suggested that tin radicals, from tributyltin hydride, might be suitable. There were already many examples of the use of tin hydrides in the reduction of bromides and iodides.[7]

So, I came to the following invention.[8] The secondary alcohol **10** would be converted to its thionobenzoate **11**. Reaction of the latter with tributyltin radicals would afford intermediate radical **12**. At a suitable temperature, **12** should fragment (Scheme 1) into a benzoylthiotin derivative **13** and the

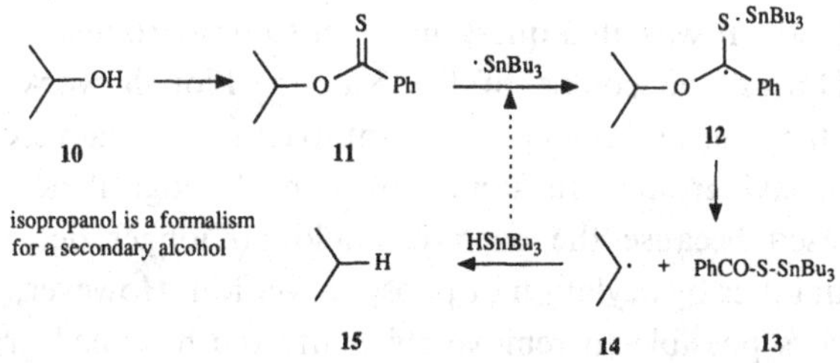

isopropanol is a formalism for a secondary alcohol

Scheme 1

secondary radical **14**. The latter would be quenched to give the desired deoxy-compound **15** and reform the tributyltin radical.

A very talented collaborator, Dr S. W. McCombie, quickly developed an improved Vilsmeier procedure for the synthesis of the desired thionobenzoate **11** and then showed that **11** was smoothly reduced to the desired deoxy-compound **15** at 80° under reflux in benzene using a suitable initiator. Our first example was with cholesterol, but a further example with the glucose derivative **16** showed the power of the new reaction. Derivative **16** gave the thionobenzoate **17** without difficulty. Reduction with tributyltin hydride under the same conditions as above gave a high yield of the desired deoxy-compound **18**.

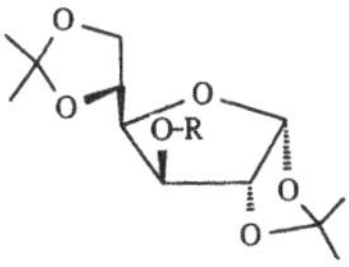

16 R = H

17 R = C(S)-Ph

19 R = C(S)-SMe

20 R = C(S)-N(imidazolyl)

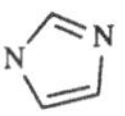

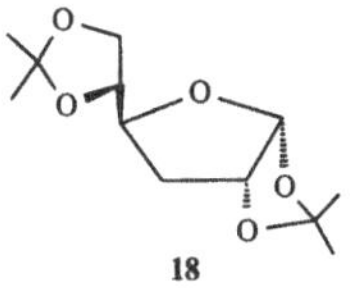

18

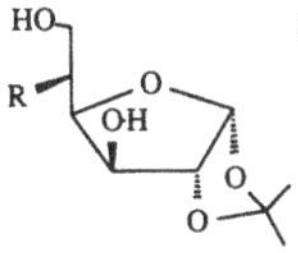

21a R = OH

21b R = H

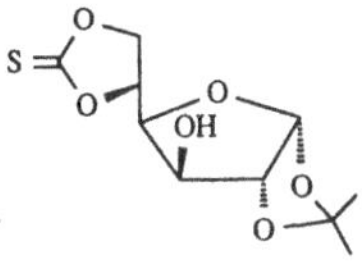

22

This was an impressive demonstration of the value of radical chemistry. Deoxygenation of **16** had never been accomplished before. Also, there was no sign of elimination products or of rearrangements.

We explored other thiocarbonyl derivatives including xanthates (e.g. **19**), thionoimidazolides (e.g. **20**) and thiocarbonates. The last convert a glycol such as **21a** via the thiocarbonate **22** and subsequent tin hydride reduction to mainly the primary alcohol **21b**. The intermediate opens in such a way as to afford the more stable radical. This reaction is readily applied to nucleosides.[9,10]

Since it seemed likely that a xanthate function beta to a carbon radical would eliminate to furnish a double bond, we studied the tin hydride reduction of 1,2-dixanthates.[11,12] This provided an efficient new synthesis of olefins which was independent of the stereochemistry of the glycol. It was particularly suitable for the conversion of protected ribosides **23** via the dixanthates **24** into the corresponding olefins **25** from which biologically important dideoxynucleosides **26** were readily obtained by hydrogenation and deprotection.[13]

In connection with the chemistry of aminoglycoside antibiotics, it was important to have a radical reaction that

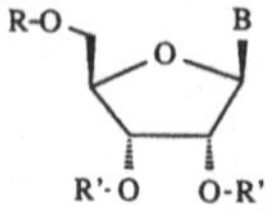

R-O O B R'-O O-R'

23 R = *t* - $BuMe_2Si$
R' = H
B = Base

24 R = *t* - $BuMe_2Si$
R' = (C=S)-SMe
B = Base

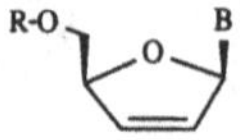

R-O O B

25 R = *t* - $BuMe_2Si$
B = Base

R-O O B

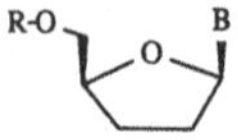

26 B = Base

deaminated in the same efficient way that the deoxygenation reaction worked. Although an earlier report had examined the tin hydride reduction of isonitriles, the results were not encouraging. However, with a suitable protected derivative of glucosamine, we found that the isonitrile derivative was converted in very high yield into the 2-deoxyglucose analogue. A general application to amino-sugar chemistry followed.[14–16]

Although tin hydride chemistry works very well in deoxygenation and in deamination reactions, the process has certain disadvantages. Tributyl- or triphenyltin hydrides have high molecular weights and are expensive for each hydrogen delivered. In addition, tin residues (usually ditin species) are always formed which are non-polar and difficult to remove from non-polar reaction products. Another disadvantage is that the quality of commercially available tin hydrides is variable.

For all these reasons, other sources for facile hydrogen atom transfer have been actively sought. Recently, tris(trimethylsilyl)silane has been shown to be an excellent reagent for this purpose.[17] It is, however, of high molecular weight and very expensive per hydrogen transferred. Our own solution to the problem is to use a phenylsubstituted silane and Ph_3SiH, Ph_2SiH_2 and $PhSiH_3$ are all suitable.[18] Triethylsilane also showed some promising results. The question of solvent participation in the hydrogen atom transfer reaction has been addressed employing fully deuterated (D_8) toluene or deuterated silanes, and the findings to our satisfaction do not show any significant C–H participation.[19]

The reaction of triethylboron with oxygen is a convenient way to generate ethyl radicals at room temperature, or at lower temperatures. These conditions at room temperature in the presence of tributyltin hydride efficiently perform the Barton–McCombie deoxygenation reaction. Recently, we showed that the tin hydride could be replaced by diphenylsilane also at room temperature, for the deoxygenation of secondary alcohols.[20] Many different thiocarbonyl derivatives can now be used for the Barton–McCombie reaction. Data were collected for R′ = Ph, *p*-F-C_6H_4, C_6F_5, $C_6Cl_3H_2$ with various substrates (Scheme 2). The yields were excellent for all the secondary

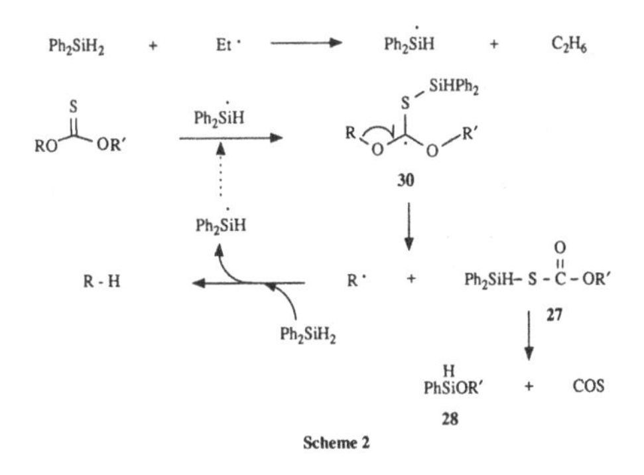

Scheme 2

alcohols. From silicon NMR measurements, the intermediate **27** as well as the final product **28** could easily be detected.

Primary alcohols, at room temperature, gave mainly the corresponding thionoformates **29**. These are surely formed from hydrogen atom quenching of the intermediate radical **30** to give **31**. This eliminates in a Peterson-like manner to give the thionoformate **29** and silane **28**, Scheme 3. This could become, at lower temperature, a useful way to make thioformates. However, if temperature is sufficiently raised then the desired

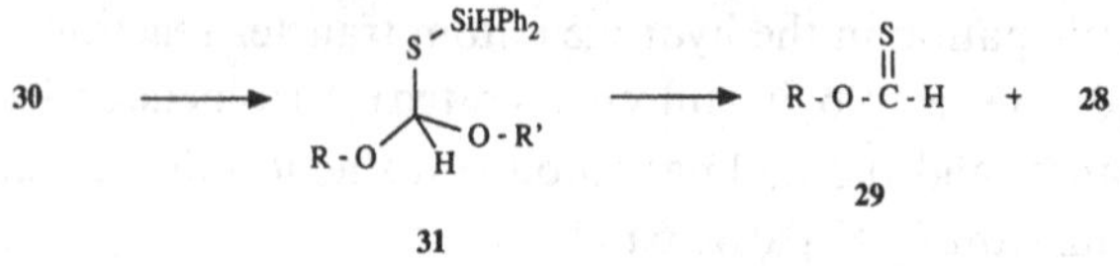

30 → **31** → **29** + **28**

Scheme 3

radical reaction does take place. Yields up to 91% in the otherwise difficult case of neopentyl type alcohol hederagenin **32** demonstrates the versatility of the method.[21]

Recent work has shown that phenylsilane, which on a mole basis is cheaper than Ph_2SiH_2, is also an excellent reagent for the Barton–McCombie reaction. Both primary and secondary alcohols are deoxygenated in very high yield in refluxing toluene using $PhSiH_3$ and xanthates or various thionocarbonates. Initiation was by azo*bis*isobutyronitrile (AIBN) or by benzoyl peroxide. The latter works well in refluxing toluene with thionocarbonyl compounds. We consider that all three phenylsilane reagents act according to Scheme 1.

Similarly the dixanthate olefin synthesis, which works well with tin hydride, can also be performed with Ph_2SiH_2.[19] Reduction of dixanthate **24** (B = adenosine) in refluxing toluene afforded **25** (B = adenosine) in 95% yield. Initiation was by AIBN or benzoyl peroxide.

The Julia olefin synthesis consists of the reaction of a sulfonyl anion with an aldehyde or ketone. The resulting alcohol is usually acetylated and the olefin formed by aluminum-amalgam reduction. The yield of β-hydroxysulfone is usually good, but the subsequent reductive elimination is more

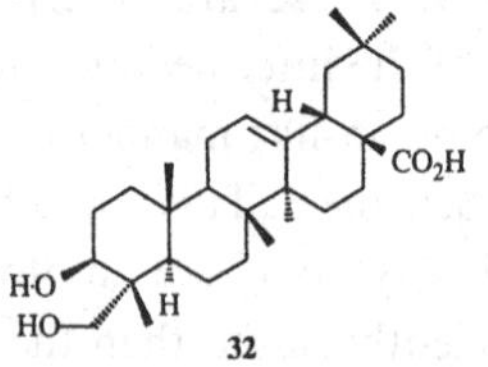

32

difficult and may give inferior yields. At the end of a long and elegant synthesis, this can be a disaster.

We conceived that, if the alcohol was to be deoxygenated to a radical, an olefinic linkage would be formed by β-elimination of a phenylsulfonyl radical. The xanthate function is a convenient radical source and gave the olefin in about 80% yield in toluene under reflux using AIBN or benzoyl peroxide as initiators and Ph_2SiH_2.[22]

References

1. Barton, D. H. R. & Miller, E. *J. Am. Chem. Soc.* (1950), **72**, 1066.
2. Barton, D. H. R. & Prabhakar, S. *J. Chem. Soc., Perkin Trans. 1* (1974), 781.
3. Barton, D. H. R., Bolton, M., Magnus, P. D., West, P. J., Porter, G. & Wirz, J. *J. Chem. Soc., Chem. Commun.* (1972), 632.
4. Achmatowicz, S., Barton, D. H. R., Magnus, P. D., West, P. J. & Poulton, G. A. *J. Chem. Soc., Perkin Trans. 1* (1973), 1567.
5. Barton, D. H. R., Chavis, C., Kaloustian, M. K., Magnus, P. D., Poulton, G. A. & West, P. J. *J. Chem. Soc., Perkin Trans. 1* (1973), 1571.
6. Barton, D. H. R., Bolton, M., Magnus, P. D., Marathe, K. G., Poulton, G. A. & West, P. J. *J. Chem. Soc., Perkin Trans. 1* (1973), 1574.
7. a) Neumann, W. P. *The Organic Chemistry of Tin*, J. Wiley, New York, 1970. b) Poller, R. C. *The Chemistry of Organotin Compounds*, Logos Press Limited, London, 1970. c) Pereyre, M., Quintard, J.-P. & Rahm, A. *Tin in Organic Synthesis*, Butterworths, London, 1987.
8. Barton, D. H. R. & McCombie, S. W. *J. Chem. Soc., Perkin Trans. 1* (1975), 1574.
9. Barton, D. H. R. & Subramanian, R. *J. Chem. Soc., Chem. Commun.* (1976), 867.
10. Barton, D. H. R. & Subramanian, R. *J. Chem. Soc., Perkin Trans. 1* (1977), 1719.
11. Barrett, A. G. M., Barton, D. H. R., Bielski, R. & McCombie,

S. W. *J. Chem. Soc., Chem. Commun.* (1977), 866.

12. Barton, D. H. R., Barrett, A. G. M. & Bielski, R. *J. Chem. Soc., Perkin Trans. 1* (1979), 2378.
13. Chu, C. K., Bhadti, V. S., Doboszewski, B., Gu, Z. P., Kosugi, Y., Pullaiah, K. C. & Van Roey, P. J. *J. Org. Chem.* (1989), **54**, 2217.
14. Barton, D. H. R., Bringmann, G., Lamotte, G., Hay-Motherwell, R. S. & Motherwell, W. B. *Tetrahedron Lett.* (1979), **24**, 2291.
15. Barton, D. H. R., Bringmann, G., Lamotte, G., Hay-Motherwell, R. S., Motherwell, W. B. & Porter, A. E. A. *J. Chem. Soc., Perkin Trans. 1* (1980), 2657.
16. Barton, D. H. R., Bringmann, G. & Motherwell, W. B. *J. Chem. Soc., Perkin Trans. 1* (1980), 2665.
17. a) Kanabus-Kaminska, J. M., Hawari, J. A., Griller, D. & Chatgilialoglu, C. *J. Am. Chem. Soc.* (1987), **109**, 5267. b) Chatgilialoglu, C., Griller, D. & Lesage, M. *J. Org. Chem.* (1988), **53**, 3641. c) Chatgilialoglu, C., Griller, D. & Lesage, M. *Tetrahedron Lett.* (1989), **30**, 2733. d) Kulicke, K. J. & Giese, B. *Syn. Lett.* (1990), 91. e) Chatgilialoglu, C., Guerrini, A. & Seconi, G. *Syn. Lett.* (1990), 219. f) Lesage, M., Martinho-Simoes, J. A. & Griller, D. *J. Org. Chem.* (1990), **55**, 5413. g) Schummer, D. & Höfle, G. *Syn. Lett.* (1990), 705. h) Ballestri, M., Chatgilialoglu, C., Clark, K. B., Griller, D., Giese, B. & Kopping, B. *J. Org. Chem.* (1991), **56**, 678.
18. Barton, D. H. R., Jang, D. O. & Jaszberenyi, J. C. *Tetrahedron Lett.* (1990), **31**, 4681.
19. Barton, D. H. R., Jang, D. O. & Jaszberenyi, J. C. *Tetrahedron Lett.* 1991, **32**, 7187.
20. a) Barton, D. H. R., Jang, D. O. & Jaszberenyi, J. C. *Tetrahedron Lett.* (1991), **32**, 2569. b) Barton, D. H. R., Jang, D. O. & Jaszberenyi, J. C. *Syn. Lett.* (1991), 435.
21. Barton, D. H. R., Blundell, P., Dorchak, J., Jang, D. O. & Jaszberenyi, J. C. *Tetrahedron* 1991, **47**, 8969.
22. a) Barton, D. H. R., Jaszberenyi, J. C. & Tachdjian, C. *Tetrahedron Lett.* (1991), **32**, 2703. See also, b) Lythgoe, B. &

Waterhouse, I. *Tetrahedron Lett.* (1977), 4223. c) Williams, D. R., Moore, J. L. & Yamada, M. *J. Org. Chem.* (1986), **51**, 3916. d) Barrish, J. C., Lee, H. L., Mitt, T., Pizzolato, G., Baggiolini, G. & Uskokovič, M. R. *J. Org. Chem.* (1988), **53**, 4282.

5

Disciplined radicals and disciplinary radical reactions (Barton decarboxylation)

SHYAMAL I. PAREKH in collaboration with
DEREK H. R. BARTON

All compounds of the arachidonic acid cascade as well as many peptides and biotin contain the carboxyl function. After seeing the value of the Barton–McCombie reaction, it was logical to consider if similar chemistry could be carried out with the carboxyl function.[1] In particular, decarboxylation to the corresponding radical seemed a promising way to replace $-CO_2H$ by $-H$.

The most useful reaction in the literature for this is the pyrolysis of a suitable per-ester in the presence of a hydrogen atom transfer reagent, but the yields are often unsatisfactory.[2,3] Conversion of an acid to the corresponding aldehyde and subsequent rhodium-based decarbonylation involves two steps, but is more reliable.[4] The Borodin–Hunsdiecker reaction converts the acid to a nor-halide, which can be reduced by radical methods. However, this works well only with primary acids, is incompatible with many sensitive functions,[5] and is expensive since the Ag salt of the acid is usually used.

Normally, esters beta to a radical are perfectly stable and do not furnish olefins and carboxyl radicals.[6] This is true even for **1** where tin hydride reduction affords only **2**. Clearly, conjugation with two phenyls is not a sufficient driving force for elimination. However, the driving force in the change from dihydrophenanthrene to phenanthrene would be greater and in

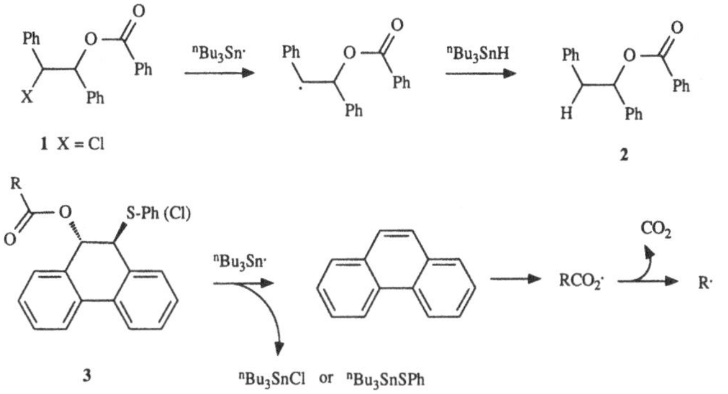

Scheme 1

fact **3** gave phenanthrene and the desired carboxyl radicals, as shown in Scheme 1.[7]

The Barton–McCombie reaction involves the breakage of a strong carbon–oxygen bond. In a hydroxylamine, or its many derivatives, the nitrogen–oxygen bond is relatively weak. Hence the acyl derivatives of thiohydroxamic acids **4** should give (with a tin radical) an intermediate **4a** where the N–O bond would readily fragment to give the carboxyl radical, Scheme 2. The analogy is clearly seen in **5** giving **5a** which is the Barton–McCombie process. The acyl derivatives of thiohydroxamic acids had already been prepared in the last century, but previously only their ionic chemistry was explored.[8] These acyl derivatives are slightly electrophilic, neutral compounds, easily prepared by acylation of the strongly nucleophilic thiohydroxamic acid function. For ease of nomenclature, they will be called esters or acyl derivatives in the sequel.

The original conception of reductive decarboxylation was in terms of tin hydride chemistry, simply because of previous experience. The earlier observations,[7] in the case of dihydrophenanthrene esters, led us to think of incorporating the C=N double bond to be formed (Scheme 2) by fragmentation, into an aromatic system. This was to provide an additional driving

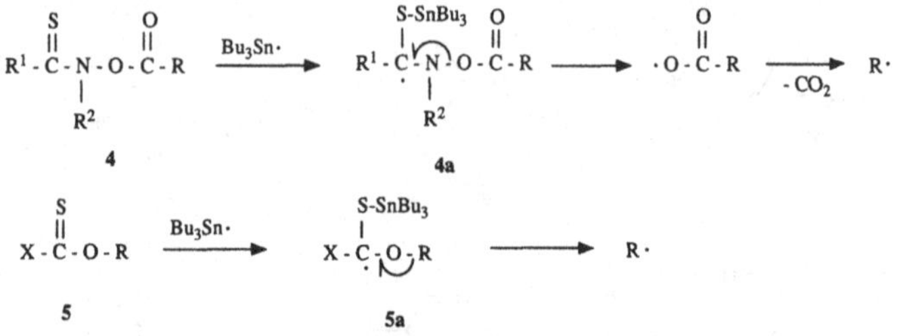

Scheme 2

force. After preliminary experiments, the esters **6**, which are derived from *N*-hydroxy-2-thiopyridone **7a**, were studied. The latter exists in partial equilibrium with its tautomer **7b** 2-thiopyridine-*N*-oxide, but the derived esters are entirely in the form shown, **6**. The normal preparations of thiohydroxamic acids are somewhat odorous. It was due to the diligence and initiative of David Crich that we became aware of the commercial existence of the sodium salt of *N*-hydroxy-2-thiopyridone **8**.[9] This salt was available under a misleading name from Fluka.[9] To make the situation even more curious, the compound was not mentioned at all in the index of the catalogue!

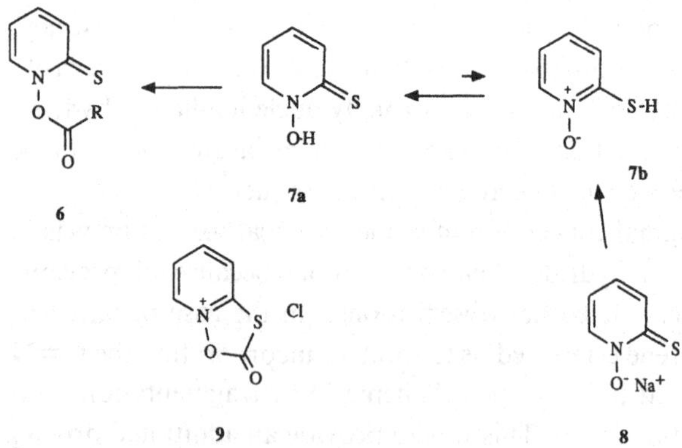

In reality, however, this sodium salt **8** is available very inexpensively from the Olin Corporation as a 40% aqueous solution. One only has to acidify the aqueous solution with concentrated hydrochloric acid and the highly crystalline thiohydroxamic acid precipitates. The requisite *o*-acyl thiohydroxamates for decarboxylation are readily prepared by esterification with an acid chloride and the sodium salt **8**. Alternative preparations involve coupling the acid **7a** and a carboxylic acid by means of a dehydrating agent like dicyclohexylcarbodiimide (DCC). The triethyl ammonium salts of carboxylic acids, upon treatment with the phosgene salt of the 2-thiopyridone-*N*-oxide **9**, yield quantitatively the desired thiohydroxamates. Other methods of preparation have been developed in recent years, and a wide range of esters **6** have

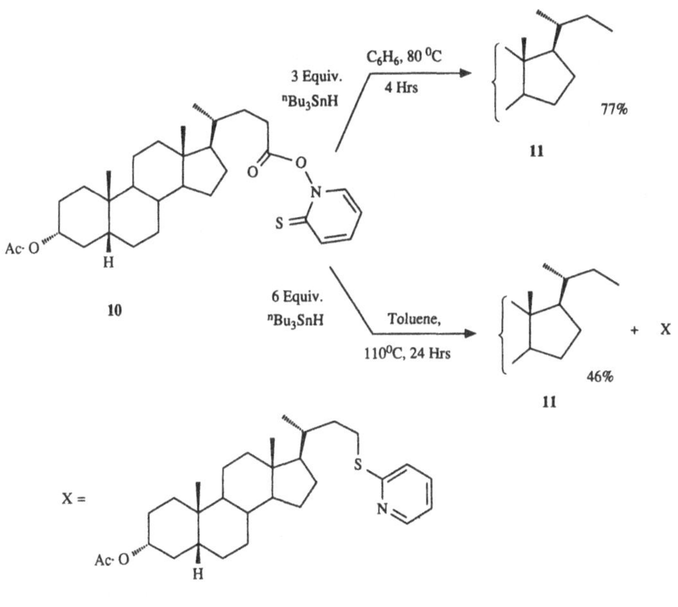

3 Equiv. $^{n}Bu_3SnH$ — C_6H_6, 80 °C, 4 Hrs → **11** 77%

10

6 Equiv. $^{n}Bu_3SnH$ — Toluene, 110 °C, 24 Hrs → **11** 46% + X

X = **12**

Scheme 3

been prepared by various groups using any one of the above-mentioned methods.[10,11]

As shown in Scheme 3, the reductive decarboxylation of **10** was carried out with tributyltin hydride in the presence of azo*bis*isobutyronitrile (AIBN) as an initiator. The results were impressive.[12] Acyl derivatives **10** underwent smooth decarboxylation to give the corresponding noralkane **11** in high yield. For primary acids, however, we were surprised to find that reactions were faster and gave higher yields in benzene at 80 °C than in toluene at 110 °C. Careful examination of the reaction mixture in the case of **10** → **11** (Scheme 3) showed the presence of another compound X, identified as the thioether **12**. Later it was concluded that **12** is the species reduced by the stannane to the noralkane.[13] Apparently, at the higher temperature, the formation of sulfide **12** competes with the desired reduction process. We found that, indeed, in the absence of the reducing agent, alkyl-2-pyridyl sulfide **12** is the exclusive product. We now had a new and mild way of generating carbon-centered radicals from the corresponding carboxylic acids.[14]

A variety of aliphatic and alicyclic acids were successfully employed to establish the scope and limitations of this new

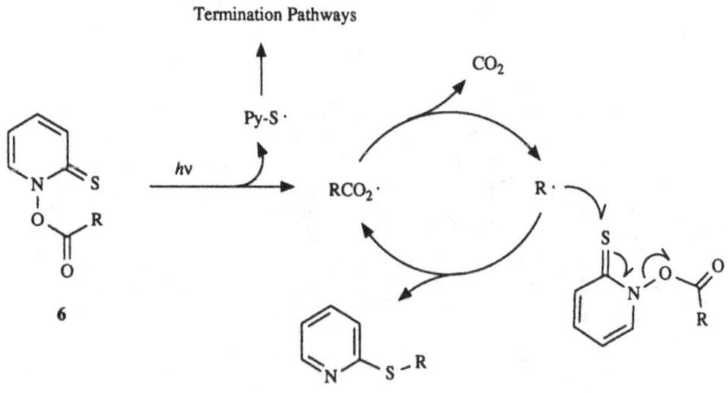

Termination Pathways

CO_2

Py-S ·

hv

RCO_2 ·

R ·

6

Scheme 4

reaction.[12–15] At this point, a radical chain mechanism was put forward to explain this novel high yielding decarboxylative rearrangement reaction, Scheme 4.[14,15] The observation that this reaction can be initiated by an ordinary tungsten lamp (visible light) was also quite remarkable.[12] Upon irradiation with a tungsten lamp, thiohydroxamates **6** undergo smooth decarboxylation to give free radicals. This has certainly opened up new dimensions of research in the area of free radical chemistry.[11] As a matter of fact it has been shown that the benzyl radicals generated via this method in a flow reactor gave ESR spectra of the highest quality.[16] Formation of the alkyl-2-pyridyl sulfide (now referred to as the background rearrangement) is the most elementary reaction that can take place photochemically at ambient temperatures,[10,11] and this observation was an important one. It suggested that the radical $R^{\cdot}$ could be captured by other radical traps. Previous experience had taught us that maximum advantage of thiocarbonyl groups and their radical trapping capabilities is taken when the trapping step is designed to be one step in a chain sequence. This enables the radical concentration to be maintained at a minimum level and thus effectively eliminates wasteful and unwanted side reactions like radical–radical coupling, etc.

This foundation led to the design of a plethora of trapping agents for a one pot transformation of carboxylic acids into various functional groups. Addition of a thiol (PhSH, *t*-BuSH, etc.) was the first such reagent employed.[12] Needless to say that it worked quite well. As shown in Scheme 5 the presence of mercaptan in slight excess gave the reductive decarboxylation product (R–H) in excellent yield. Besides the regular aqueous workup, and a quick filtration over silicagel, the product mixture did not require any further purification.[12,15,17] The necessity for expensive and toxic tin reagents was thus eliminated.

At this stage any skepticism invoking a two-electron (intramolecular) concerted mechanism to explain the formation

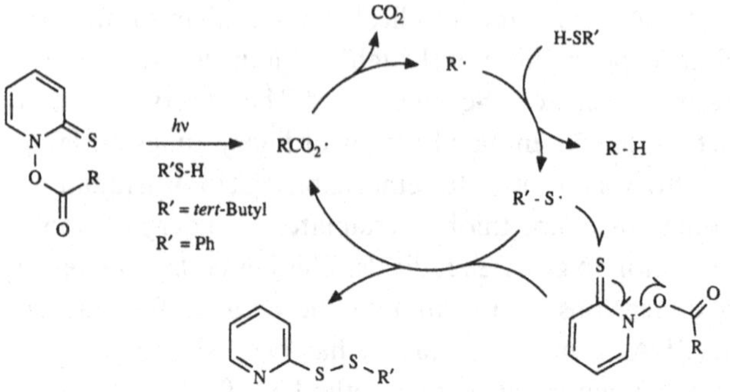

Scheme 5

of alkylpyridylsulfide could be easily eliminated. Stereoelectronic considerations disfavor this concerted mechanism as well.[18] The thiol itself could not possibly reduce the alkylsulfide to the alkane, hence free radicals had to be involved. Two different radical mechanisms can be written for this reaction. The alternative to an obvious free radical chain reaction is the radical cage (leaky!) mechanism involving homolytic cleavage of the N–O bond (Scheme 6). Such caged radicals have previously been investigated quite extensively.[19] So the question of free/caged radicals was answered in crossover experiments.[20]

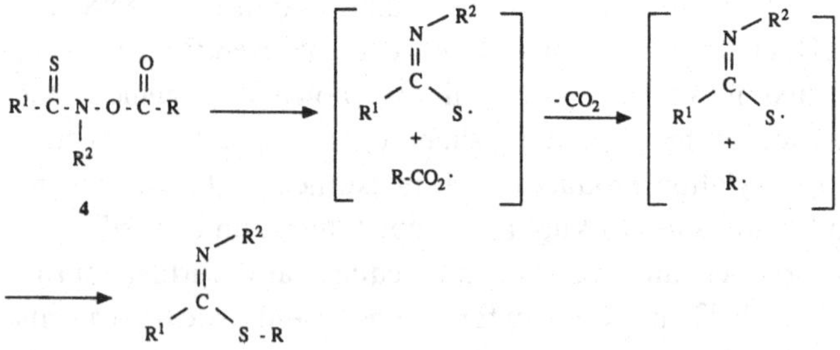

Scheme 6

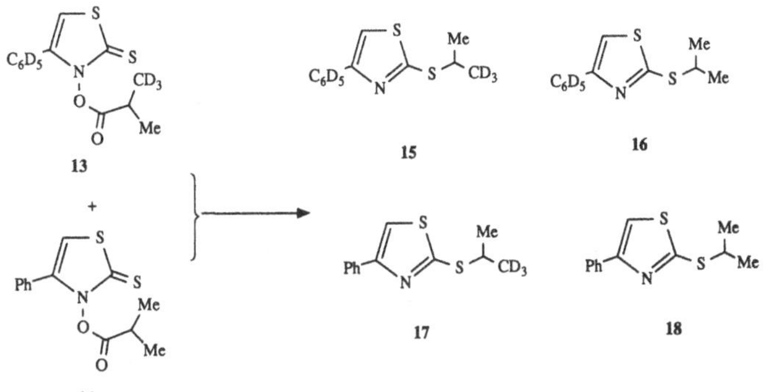

Scheme 7a

As shown in Scheme 7a, mixtures of esters **13** and **14** in equal amounts were subjected to the thermal and photolytic decarboxylative rearrangement reactions. There was a complete scrambling of both alkyl moieties in the photochemical reaction giving **15**, **16**, **17** and **18** in equal amounts.[20] The thermal process, on the other hand, gave 80 % scrambling and the remaining 20 % of the products were found to be formed due to 'caged radicals'. Further support for the proposed mechanism, and the answer to a possibility of the reversible nature of addition of R$^{\bullet}$ to the thione, was provided in a series of temperature dependence studies.[21] Zard suggested that the process is reversible as shown in Scheme 7b, and this is a very crucial point from the kinetic considerations. Relative values of k_1, k_{-1} and k_2 dictate the synthetic utility and versatility of the radical chemistry of these thiohydroxamic acid derivatives. The

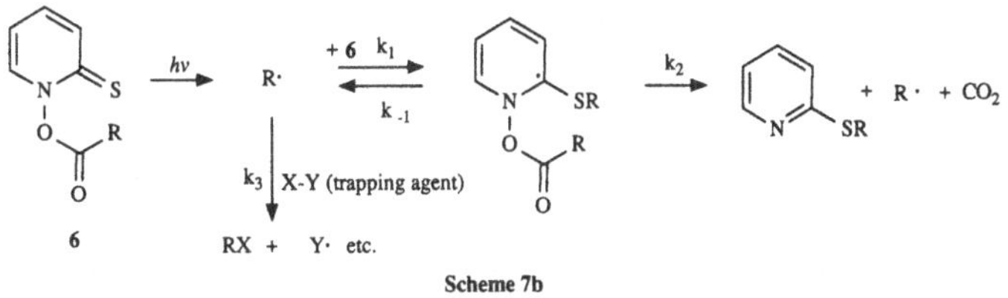

Scheme 7b

pyridinyl radical adduct can undergo two unimolecular fragmentations but k_{-1} is reversible and k_2 is irreversible so the temperature becomes an important criterion. Both processes are very sensitive to temperature variation and to the nature of R. With primary alkyl radicals, both unimolecular processes are found to be slower compared to the secondary and tertiary radical species. At higher temperatures, however, k_2 becomes greater than k_3 (competing reaction) and thus the background rearrangement predominates. Ingold, Newcomb, Beckwith and others have measured the rate constants of this reaction and other radical chain processes. They have clearly shown that the limiting rates of scavenging of the radicals ($k_{\text{competing reaction}}$) are the determining factors in the overall efficiency of any radical chain reaction.[22] The factors influencing the lifetime of the radicals in this reaction are undoubtedly the rate of addition to the starting thiohydroxamate, the reversible nature of this addition and thus the temperature of the reaction, and the nature of trapping X–Y.

This reaction, nevertheless, has been designed with suitable trapping agents and appropriate conditions so that various rate constants fit together and allow efficient interception of the free radicals making a synthetically versatile reaction. It is clear (today) that in the radical deoxygenation reaction the radicals are disciplined by the weak Sn–H bond, whereas, in the acyl derivatives of the thiohydroxamic acid, the thiocarbonyl group is the ultimate disciplinary group. This disciplining process is aided by the inherent weakness of the N–O bond. Aromatization of the mercaptopyridine nucleus (enthalpy favored) and expulsion of CO_2 (entropy favored) are additional driving forces behind this reaction.

Now, having at hand a convenient source of carbon radicals, we considered ways of intercepting them by various reagents, thus diverting the reaction from its normal course. However, to retain all the advantages of disciplined radical reactions via the chain mechanism, the choice of an appropriate propagating radical was essential. Thiol, being an excellent hydrogen atom donor worked well as we saw earlier where the thiyl radical propagates the chain, Scheme 5. As a consequence of similar mechanistic considerations, we turned our attention to the Borodin–Hunsdiecker reaction. With the exception of the tert-butyl hypoiodite method,[23] all the other methods/modifications use heavy metal salts such as those of lead,[24a] and thallium,[24c] and the disadvantages of such procedures are obvious.[24,25] We conceived that the decomposition of **6** in suitable halogen atom sources would result in the formation of nor-alkyl halides in a radical chain reaction, Scheme 8a.

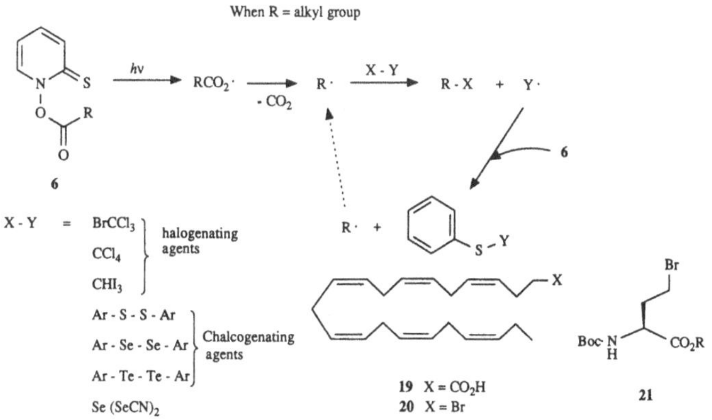

When R = alkyl group

X - Y = $BrCCl_3$, CCl_4, CHI_3 (halogenating agents); Ar - S - S - Ar, Ar - Se - Se - Ar, Ar - Te - Te - Ar (Chalcogenating agents); Se $(SeCN)_2$

19 X = CO_2H
20 X = Br

21

Scheme 8a

Needless to say, this was readily proven to be the case and a variety of chlorides, bromides and iodides have been prepared from the corresponding acids in good to excellent yields, by this method.[14,15,26–31] In terms of mildness of conditions, generality and yields, this method is far superior to the classical Borodin–Hunsdiecker reaction and its variants.[10,11,26–31] As a demonstration of the mildness of the conditions, decarboxylative bromination of the heavily functionalized acid **19** was carried out by Professor Ikegami to afford **20** (75 % yield), and it is important to point out that all the other classical methods had failed on previous attempts in this case.[32] This method has been applied to primary, secondary and tertiary acids, including a desirable formation of bromide **21**, derived from the corresponding glutamic acid in a one pot conversion.[17] When this method was extended, however, to aromatic and α,β-unsaturated carboxylic acids, the results were initially disappointing. Upon irradiation with light, the acyl derivatives of thiohydroxamates afforded the acid and the anhydride as

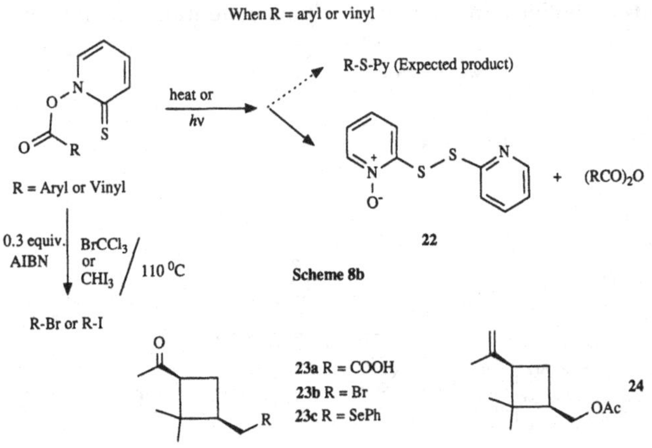

When R = aryl or vinyl

R-S-Py (Expected product)

heat or $h\nu$

R = Aryl or Vinyl

+ $(RCO)_2O$

0.3 equiv. AIBN

$BrCCl_3$ or CHI_3 / 110 °C

22

Scheme 8b

R-Br or R-I

23a R = COOH
23b R = Br
23c R = SePh

24

OAc

shown in Scheme 8b, and the expected sulfide formed to a very small extent and pyridyl disulfide mono-oxide **22** was the major product.[33a] Professor Vogel suggested increasing the effective concentration of $^{\bullet}CCl_3$ by incorporating AIBN into the reaction mixture.[33b] In effect this was to facilitate the fragmentation of the first adduct radical. A mixture of acid chloride, sodium salt of *N*-hydroxy-2-thiopyridone and AIBN (30 %) in hot (110 °C) $BrCCl_3$ caused a marked improvement in the overall yield of the product. A variety of aromatic acids as well as cinnamic acid could be transformed cleanly into the corresponding bromo and iodo derivatives in reasonable to excellent yields.[33a,34] This application is particularly useful since electron-rich aryl acids suffer extensive electrophilic aromatic bromination when their silver salts are heated with bromine (the normal Borodin–Hunsdiecker conditions).[10]

Reductive chalcogenation was also achieved using disulfides, diselenides and ditellurides.[15,35,36] Dicyanogen triselenide was found to be a good trapping agent and a chain carrying agent as well in this novel radical chain reaction, Scheme 8a.[37] An interesting application of this decarboxylative chalcogenation and halogenation was in the case of (+)-*cis*-pinonic acid **23a** (readily available from α-pinene). This was accomplished in connection with the synthesis of the acetate **24**, the sex pheromone of the citrus mealybug.[38]

The functional group transformation $R-CO_2H \rightarrow R-OH$ is normally a multistep procedure (e.g. $R-CO_2H \rightarrow R-Br \rightarrow R-O-C(O)R' \rightarrow R-OH$). Triplet oxygen is an excellent radico-

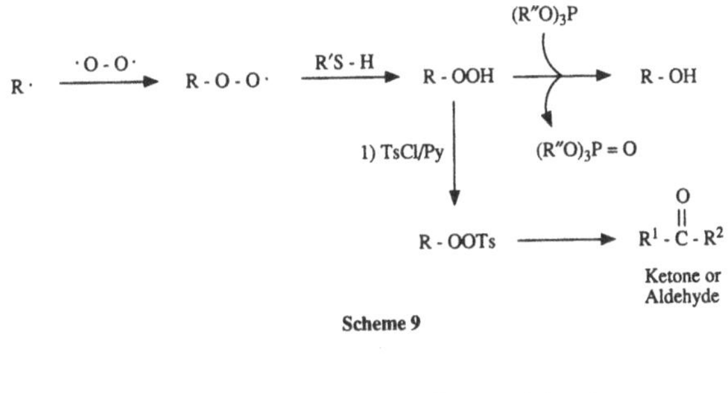

$$R\cdot \xrightarrow{\cdot O-O\cdot} R-O-O\cdot \xrightarrow{R'S-H} R-OOH \xrightarrow[-(R''O)_3P=O]{(R''O)_3P} R-OH$$

$$R-OOH \xrightarrow{1)\ TsCl/Py} R-OOTs \longrightarrow R^1-\overset{O}{\overset{\|}{C}}-R^2$$

Ketone or Aldehyde

Scheme 9

phile. The reaction (Scheme 9) was performed in the presence of a thiol to reduce the intermediate hydroperoxy radical to the corresponding hydroperoxide and furnish the thiyl radical which carried the chain.[15,39] Conditions were established such that a large excess of oxygen was available compared to the thiol and hence premature reduction of the alkyl radical could be successfully avoided. It is possible to isolate these hydroperoxides if desired. However, we have found it convenient to reduce the hydroperoxide *in situ* with trialkyl phosphite to the nor-alcohol.[15] Conversion of hydroperoxides to the corresponding sulfonate derivatives *in situ* and elimination has also provided the corresponding aldehydes or ketones depending on whether the starting acid is primary or secondary.[39]

We next examined a series of transformations crucial to organic chemists, namely the formation of carbon–carbon bonds. Giese, Curran and several others have also looked at the addition of alkyl radicals to carbon–carbon multiple bonds, and it has been found that for the intermolecular addition reaction good to excellent results are obtained if the carbon–carbon multiple bond is sufficiently activated.[11,30,31,40–43] Though the relative rates of the chain sequence determine the outcome of the reaction, the frontier molecular orbital approach to this problem helps in developing better predictability for the choice of a suitably substituted olefin for the desired efficiency.[42] During the decarboxylative radical addition to carbon–carbon double bonds using thiohydroxamates, the background rearrangement reaction is a major competing side reaction and yields are variable, being highest for reactive, not easily polymerized olefins. An interesting reagent is an appropriately substituted olefin **25**, Scheme 10.[44] By acting as a leaving group, the 'X' moiety prevents polymerization and efficiently propagates the chain.[44,45] The net result is allylation of the alkyl radical, albeit in moderate yield. Free radical allylation of halides has been previously attempted and

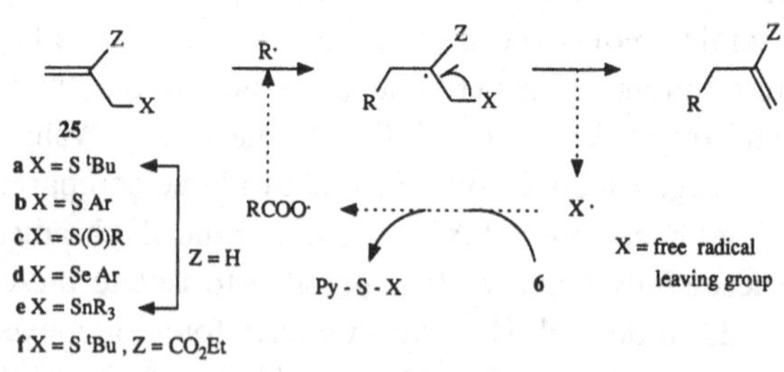

Scheme 10

successfully employed in the synthesis using allyltributyl stannane (an S_H2' reaction).[46] In our study the background rearrangement was the major competing side reaction, and further activation of the carbon–carbon double bond was necessary. Choice of **25f** improved the yields considerably giving carbethoxyallylation of the alkyl radical.[44,45]

As exemplified in Scheme 11, use of a strongly electron-withdrawing functional group (e.g. $-NO_2$) as the activating

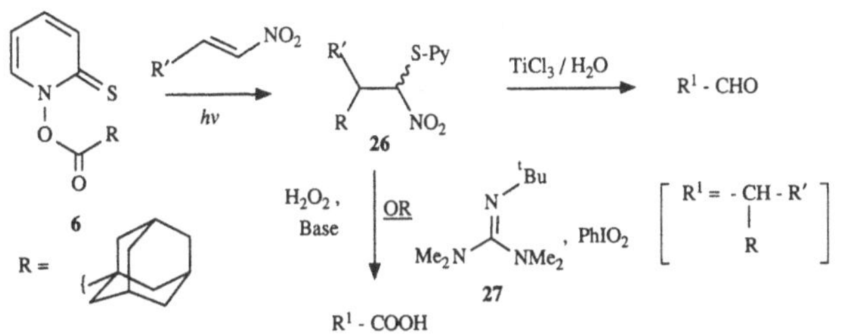

Scheme 11

group, makes radical addition even to internal alkenes a high yielding reaction.[47] The use of nitroalkenes as radical traps in this sequence is especially beneficial as the product α-nitro sulfides **26** can be easily converted to the carboxylic acids by treatment with either iodoxybenzene/tert-butyltetramethyl guanidine **27** or alkaline hydrogen peroxide, Scheme 11. With aqueous $TiCl_3$ the α-nitro sulfides yield aldehydes. Use of maleic anhydride as a radical trap provided 2-alkyl-3-aryl-thiosuccinic anhydrides **28** which, under the reaction conditions, spontaneously eliminated the thiol to give **29** in a very clean reaction, Scheme 12. This is an excellent method to make substituted maleic anhydrides.[10] Maleimide and substituted cyclopent-2-ene-1,3-dione underwent similar high yielding reactions as outlined in Scheme 12.[48]

Vinyl phosphonium salts **30** are also very efficient radical traps. As shown in Scheme 13, the corresponding adducts **31**,

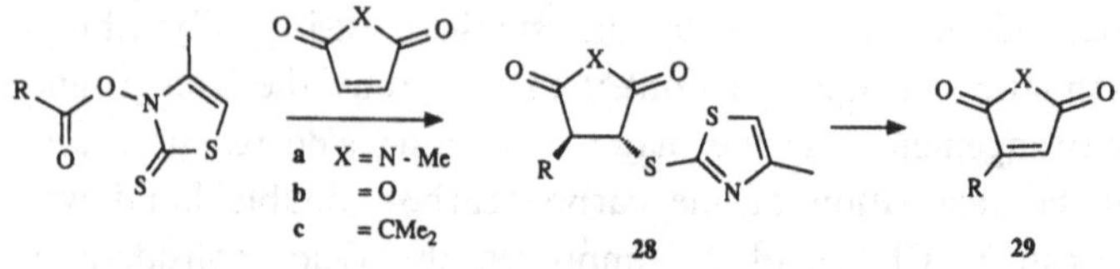

Scheme 12

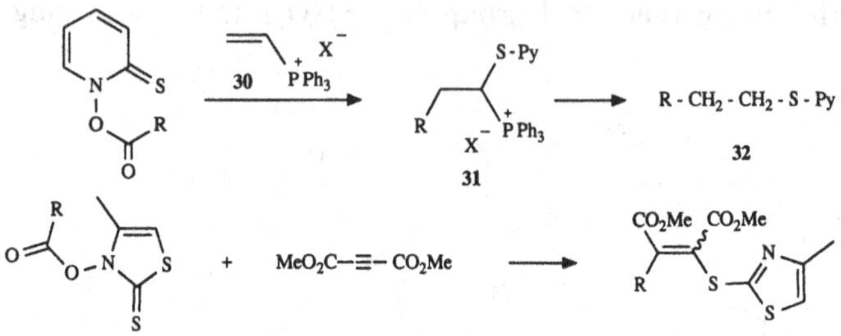

Scheme 13

upon base hydrolysis, provided the simple chain elongated alkyl pyridyl sulfide, **32**.[49] Radical addition to electron-deficient acetylenes has also been achieved in reasonable yield using this protocol.[48,50] Similarly, addition of electrophilic radicals generated from the corresponding perfluoroacyl derivative of thiohydroxamate **6** (R = $-CF_2CF_2CF_3$) to electron-rich olefins (vinyl ethers, for example) was achieved in moderate yields, Scheme 14.[51]

Free radical cyclizations for the construction of monocyclic or multicyclic ring systems have been carried out by many groups.[11,30,31,40–43] The radical chemistry based on the thione

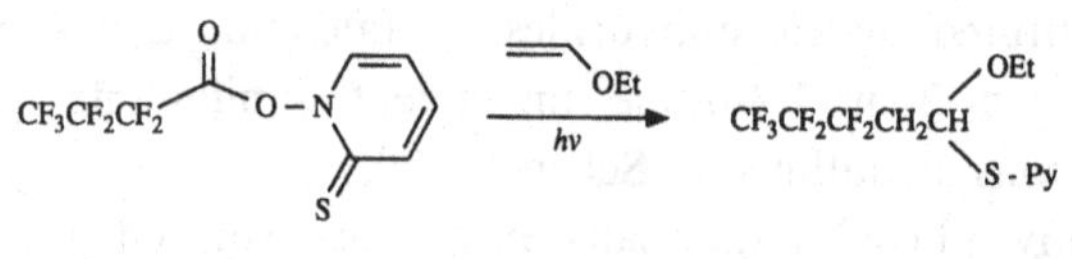

Scheme 14

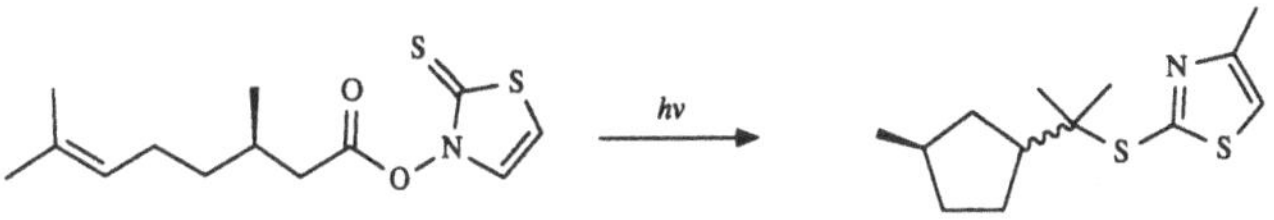

hν

Scheme 15a

CO_2H

33

$PhSO_2$

S-Py

SO_2Ph

34

Scheme 15b

X = -COMe
-SO_2Ph
-CN
-CO_2Me
-$CO_2Ad(2)$

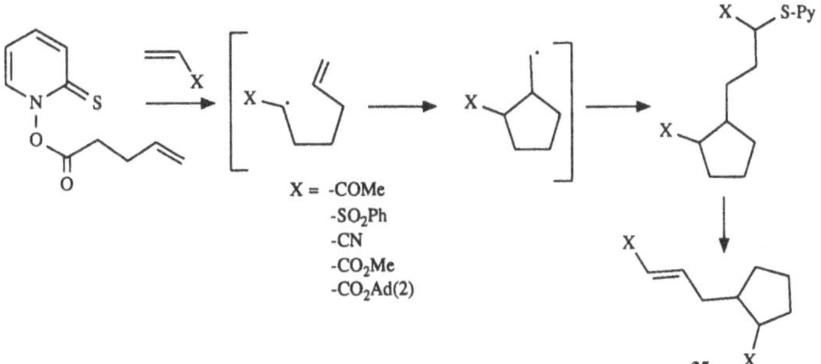

35

Scheme 15c

function offers certain distinct advantages over metal-based radical chemistry.[11,30] So, our methodology was tested in radical cyclization reactions via the intramolecular addition of a carbon radical to a suitably placed carbon–carbon double bond in the molecule as exemplified in Scheme 15a.[50,52] A somewhat different approach to the formation of carbon–carbon bonds by the radical decarboxylation method involved an addition–cyclization–addition sequence. Zard has elegantly utilized this sequence allowing the formation of complex bicyclic systems **34**, from simple γ,δ-unsaturated acids **33** and 2 equivalents of an electron-deficient alkene (phenyl vinyl sulfone) as exemplified in Scheme 15b.[53] This protocol also offers efficient conversion of readily available starting materials as outlined in Scheme 15c into a variety of substituted cyclopentane derivatives **35**, including an acyl substituted cyclo-

hv Z - M

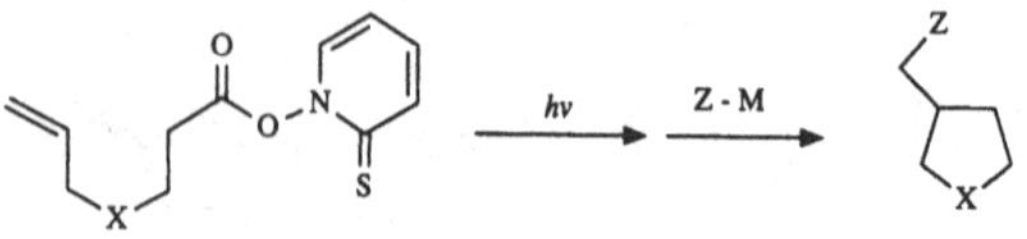

36a X = N - Ac
36b X = N - $CH_2CH{=}CH_2$
36c X = O

37

Scheme 16

pentane where methyl vinyl ketone was successfully employed as a radical trap.[54] Professor Corsano has successfully utilized this approach in a simple synthesis of pyrrolidines and tetrahydrofurans as outlined in Scheme 16.[55] Photolysis of the readily available *N*-hydroxy-2-thiopyridone derivatives of 3-*N*-acetylallylamino-, 3-*N*,*N*-diallylamino- and 3-allyloxy-propionic acids **36a**, **36b**, and **36c** was carried out, which gave cleanly derivatives of *N*-acetyl-3-methylpyrrolidine **37a**, *N*-allyl-3-methylpyrrolidine **37b** and of 3-methyl tetrahydrofuran **37c** respectively.[55]

Labelling carboxylic acids with ^{13}C or ^{14}C could be useful in biological experiments. There is no short yet convenient way of accomplishing this by classical methods in few steps. We conceived that R$^{\bullet}$ radicals generated from a carboxylic acid should be able to react with a suitable one carbon (labelled) trap, which could then be converted back into the carboxyl function. The problem was to find a sufficiently active 'one carbon reagent' which could be manipulated under mild conditions. The sort of functionality that exists in leukotrienes (for example), suggests that the reaction be carried out in neutral conditions. Our first solution to the problem was to use suitably activated isocyanides as one carbon fragments, and they worked quite well, as shown in Scheme 17.[56] In the case of some other isocyano derivatives, the stability of the isocyanide was a major limiting factor, so the isocyanides **38a** and **38b** were the best compromise between reactivity towards radicals and stability. Alkyl radicals generated upon photolysis reacted

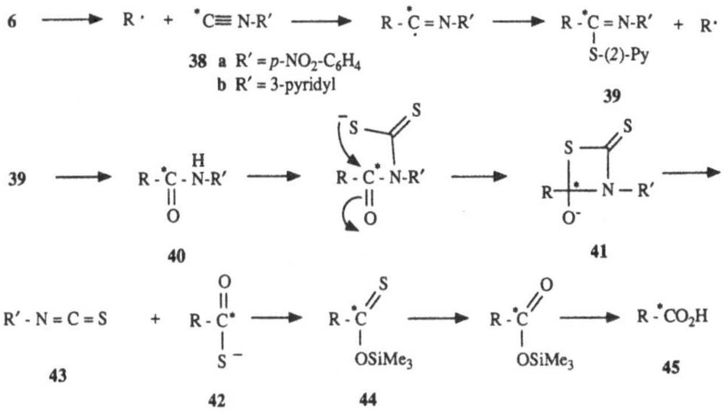

6 ⟶ R· + ·C≡N-R′ ⟶ R-Ċ=N-R′ ⟶ R-Ċ=N-R′ (S-(2)-Py) + R·

38 a R′ = *p*-NO_2-C_6H_4
b R′ = 3-pyridyl

39

39 ⟶ R-Ċ(=O)-NH-R′ (**40**) ⟶ ⟶ **41** ⟶

R′-N=C=S (**43**) + R-Ċ(=O)-S^- (**42**) ⟶ R-Ċ(=S)-$OSiMe_3$ (**44**) ⟶ R-Ċ(=O)-$OSiMe_3$ ⟶ R-ĊO_2H (**45**)

Scheme 17

with **38a** and (in the presence of trifluoroacetic acid) **38b** to give the adducts as shown in Scheme 17. Convenient reaction procedures have been worked out to hydrolyze the adducts of the type **39** to the corresponding amides. The adducts **39** can readily be hydrolysed with water to give **40**. An ingenious procedure[57] for the conversion of secondary amides to thioacids and isothiocyanates was adapted to our problem. Normally,[57] the anion, generated with sodium hydride, reacts with CS_2 to give via **41** the thioacid anion **42** and isothiocyanate **43**. We conceived that, if hexamethyldisilazane anion was used as base, the thioacid anion **42** would be silylated[58] on oxygen *in situ* to give the thiocarbonyl derivative **44**. So addition of phenylseleninic acid[59] would then convert the thiocarbonyl to carbonyl, also *in situ*, and give, on addition of water, the desired labelled carboxylic acid **45**. This idea worked well[56] and should find other applications.

The weak step in this synthesis is the radical addition to the isonitrile. Isonitriles that are sufficiently radicophilic are also easily polymerized. So we decided to develop a better procedure. The sulfonylcyanide function shows some radical behavior.[60] This, in principle, can be utilized for constructing a radical chain process. We decided to compare the well-known *p*-toluenesulfonyl cyanide with methane sulfonyl cyanide, a reagent that had not been used in radical reactions before. Neither reagent had been studied in our thiocarbonyl mediated radical chemistry.

Radicals generated by the photolysis (W light) of **6** readily reacted with the cyanide function of *p*-toluenesulfonyl cyanide to furnish (Scheme 18) the appropriate nitrile and the sulfonyl

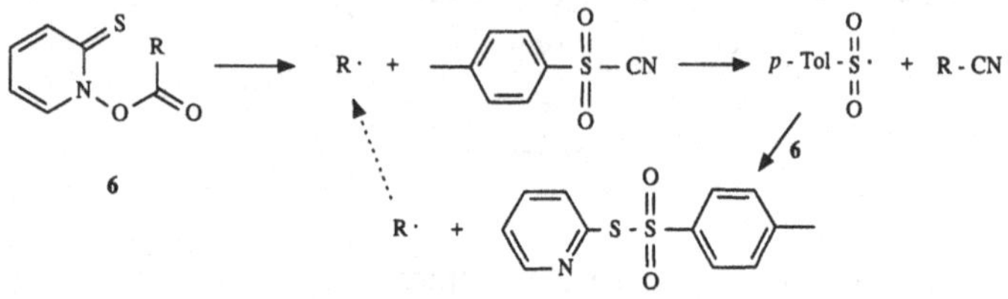

Scheme 18

radical. This reacted with the thiocarbonyl group of **6** in the usual way and reformed the R˙ radical. Methane sulfonyl cyanide turned out to be even more reactive towards carbon radicals and furnished excellent yields of R–CN.[61] Conditions for alkaline hydrolysis, which did not conjugate skipped dienes like linoleic acid, were developed. This offers an efficient synthesis of labelled carboxylic acids by simply employing the appropriately labelled nitrile moiety.

During the course of this work, we decided to examine the possibility of adding carbon radicals to diethylazodicarboxylate **46**. When **6** and **46** were left in CH_2Cl_2 at room temperature with tungsten lamp irradiation, or in the dark, they rapidly reacted to give compounds of type **47**, a class of substances never seen before.[62] On photolysis, unusual dimers **48** were produced with four linked nitrogen atoms. The formation of **47** is suggested to be as shown in Scheme 19.

Quinone and hydroquinone subunits occur in a wide variety of important natural products including the ubiquinones and

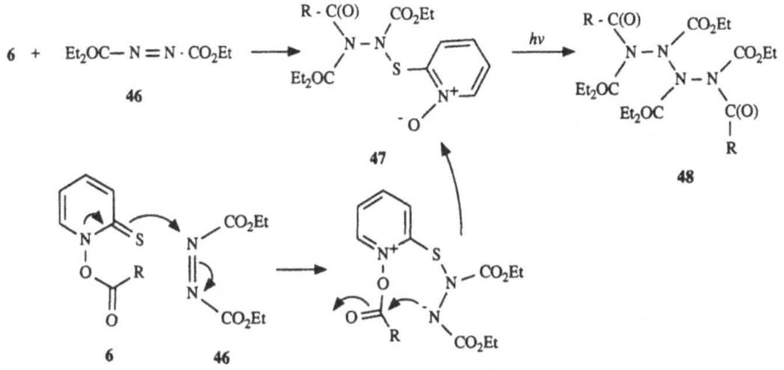

Scheme 19

vitamins E and K. A tremendous effort has been expended in search of ways for the introduction and further modification of these substructures. The synthesis of hindered quinones can be accomplished with difficulty using ionic reactions. We decided[64] to explore the limits of radical chemistry by adding tert-adamantyl radicals to quinone. The photolysis of **6** (R = tert-adamantyl) in the presence of benzoquinone **49** ($R^1 = R^2 = R^3 = R^4 =$ H) afforded the corresponding adduct **50** which on per-acid oxidation readily eliminated PySOH, to give **51** (R = tert-adamantyl, $R^1 = R^3 = R^4 =$ H). Addition of a second tert-adamantyl group to **51** afforded, after oxidative elimination, the two hindered quinones **51a** and **51b**, easily distinguished from each other by ^{13}C NMR. We have accomplished the alkyl radical addition to various other quinones **49** in high yields, Scheme 20.[63,64] Different procedures have been developed for the conversion of adducts **50** into hindered quinones of type **51**. Hydroquinones of the type **52** are also readily accessible via this protocol.[63]

The homologation of carboxylic acids is a reaction frequently needed in synthetic chemistry. The Arndt–Eistert reaction though aesthetically pleasing, is not preferred today, owing to the use of diazomethane and silver reagents. We provided[47] a solution to this problem by the addition of carbon radicals to

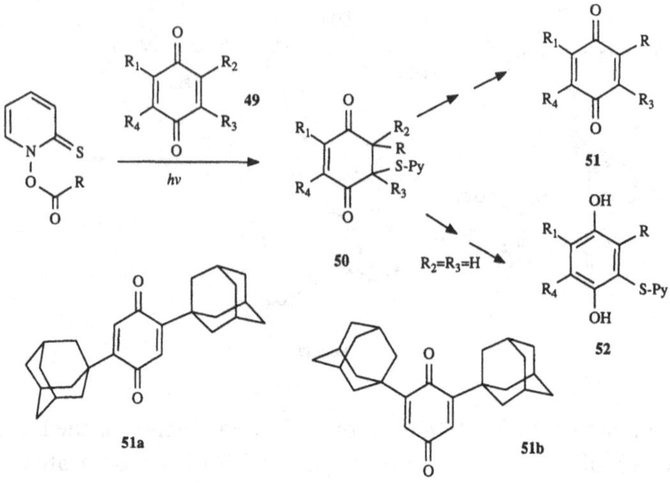

Scheme 20

nitroethylene, as discussed earlier (Scheme 11, R′ = H). The adducts **26** were converted in high yield to the homocarboxylic acids with H_2O_2 under mild basic conditions (K_2CO_3; 40°). However, it is not easy to make nitroethylene on a large scale. We have therefore looked for another solution to the problem.[65] Addition of radicals from **6** to phenylvinylsulfone **53a** is a very efficient, known reaction to give **54a**. Oxidation to sulfoxide followed by Pummerer rearrangement with trifluoroacetic anhydride affords derivatives **55**, Scheme 21. Mild alkaline hydrolysis affords **56**. The Perkow reaction[66] on phenylthiochloroacetate gave derivative **57**, easily oxidized to the sulfone **53d** [X = $-OP(O)(OMe)_2$]. Addition of R$^{\bullet}$ radical to the latter afforded **54d** which was smoothly hydrolyzed to acid **56** in the presence of aqueous base. The enol phosphate **53d** is noteworthy, as its choice as a radical acceptor eliminates the Pummerer rearrangement sequence and thus higher yields have been obtained.

The manipulation of geminal 2-*S*-pyridyl phenyl sulfones

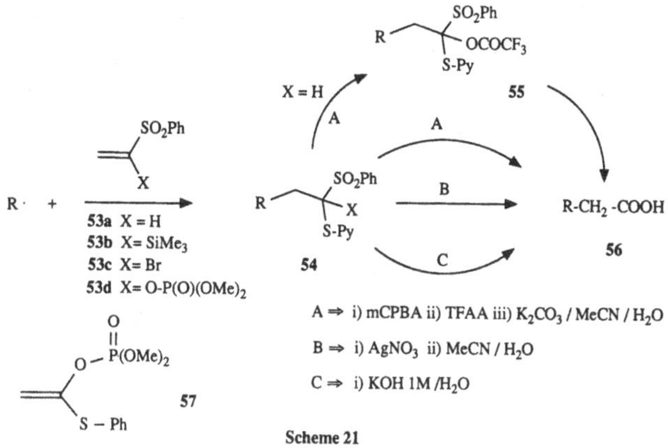

A ⇒ i) mCPBA ii) TFAA iii) K_2CO_3 / MeCN / H_2O

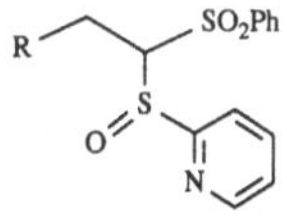

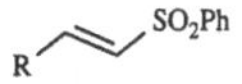

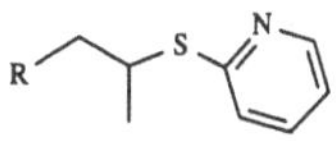

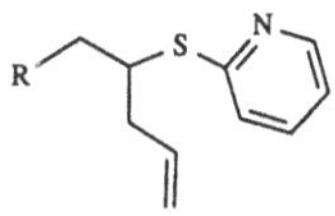

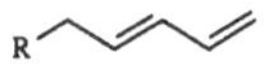

B ⇒ i) $AgNO_3$ ii) MeCN / H_2O

C ⇒ i) KOH 1M /H_2O

Scheme 21

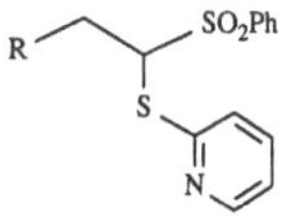

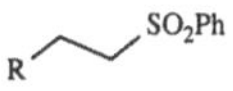

54a formed from the addition of carbon radicals R• to phenylvinylsulfone[67] was also examined from another perspective. Reduction with sodium hydrogen telluride removed the 2-*S*-pyridyl function to give **58**. Oxidation of **54a** to the sulfoxide **59** and thermal elimination gave the vinyl sulfone **60** which afforded, with sodium hydrogen telluride, the vinylic olefin **61**. The phenylsulfone group could also be removed selectively. On treatment with trimethylaluminum **54a** gave **62**, while ethylaluminum dichloride and allyltrimethyl silane afforded the allylated derivative **63**. Oxidation to the corresponding sulfoxide and thermal elimination gave **64**. All these reactions proceeded in good yield. Thus the chemistry of this geminal function, based on radical chemistry, has been considerably expanded.

The formation and photochemical decomposition of an *o*-acyl thiohydroxamate **65**, Scheme 22, which is readily

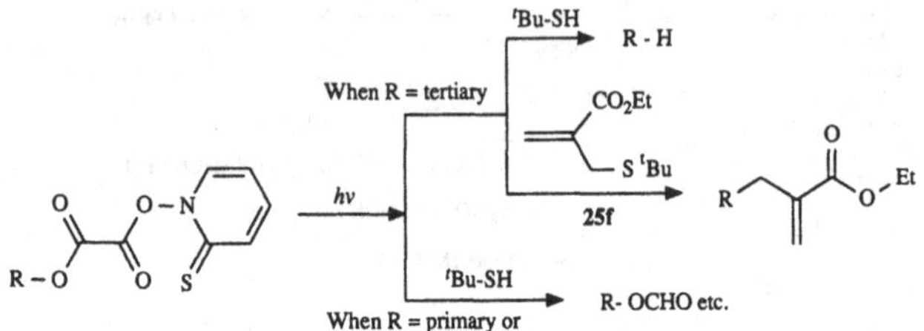

Scheme 22

available from **7** and hemioxalate esters, was investigated in the presence of a thiol. It was found that the primary and secondary alkoxycarbonyl radicals underwent slow fragmentation.[68] To our delight, however, derivatives of tertiary alcohols fragmented quite readily to give desired deoxygenated products in high yields. Considering the difficulties in preparing thiocarbonyl esters of such alcohols, the process nicely complements the Barton–McCombie procedure.[27–29, 68] Interception of the tertiary radical with the acrylate **25f** resulted in the formation of a quaternary center directly from an alcohol. The synthetic potential of this transformation is considerable given the difficulties usually encountered in creating quaternary centers.[27–29]

The masterly work of Professor Minisci has clearly shown that polar effects can play an important role in free radical reactions when charged species are involved.[69] In a series of publications, Professor Minisci and coworkers have elegantly demonstrated various aspects of these polar effects. They have shown that selective substitutions can be accomplished by reactions of nucleophilic carbon-centered radicals with electron deficient substrates. Substitution of protonated heteroaromatic bases (Het–H + R$^{\bullet}$ → Het–R) in this fashion has been developed as one of the most general reactions in heteroaromatic chemistry.[69] On similar grounds, we, in collaboration with Professor S. Corsano, studied decarboxylative radical addition onto protonated heteroaromatic systems as exemplified in Scheme 23.[70] Professor Minisci has kindly pointed out to us

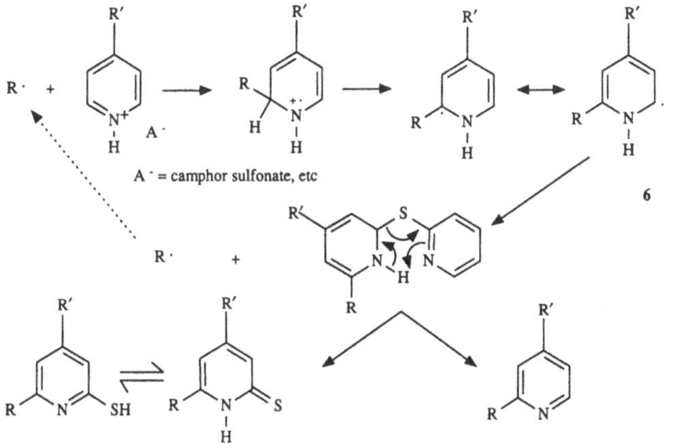

A$^-$ = camphor sulfonate, etc

Scheme 23

that, besides the rearomatization consideration to explain the product formation as shown in Scheme 23, the pyridinyl radical adduct which is a cyclohexadienyl type radical (being a highly reducing species with ionization potential similar to that of sodium) could react with **6** in an electron transfer process giving the same products. Nevertheless, it was pleasing to see that the radical chemistry of *N*-hydroxy-2-thiopyridone was found to be very compatible with an acidic medium.[70a] This prompted us to exploit this radical sequence in the syntheses of some substituted heteroaromatic systems. Camphor sulfonates of the heteroaromatic compounds were used since they are

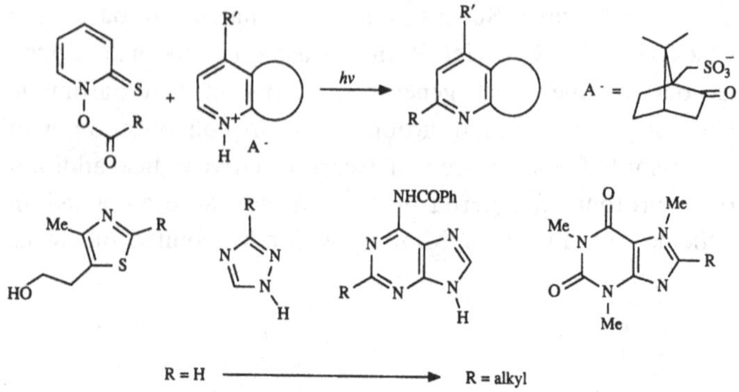

R = H → R = alkyl

Scheme 24

easier to dry and are nicely soluble in CH_2Cl_2. As shown in Scheme 24, a variety of ring systems including purines have been successfully employed. One of the interesting aspects of this free radical alkylation is the fact that it reproduces, on basic heterocyclic compounds, the equivalent of the Friedel–Crafts aromatic substitution.[69,70] This work has also allowed easy access to substituted purine bases, some of which are of considerable biological significance.[70b]

This type of chemistry is well suited for the manipulation of amino acids and peptides which tend to undergo racemization at the α carbon under ionic conditions. First examined was the decarboxylation of *N*-protected amino acids by using a thiol as hydrogen atom transfer reagent, Scheme 25.[71] The power of

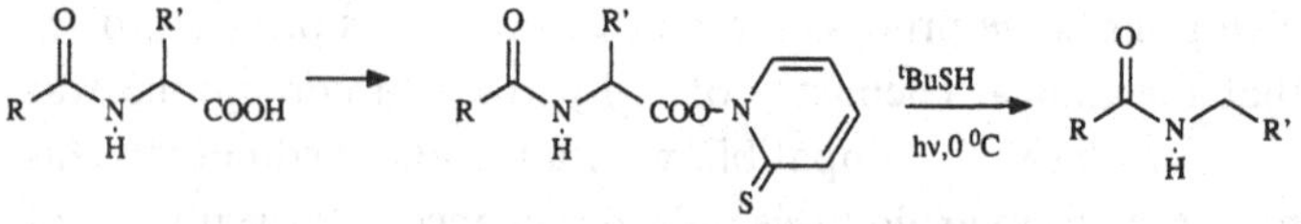

Scheme 25

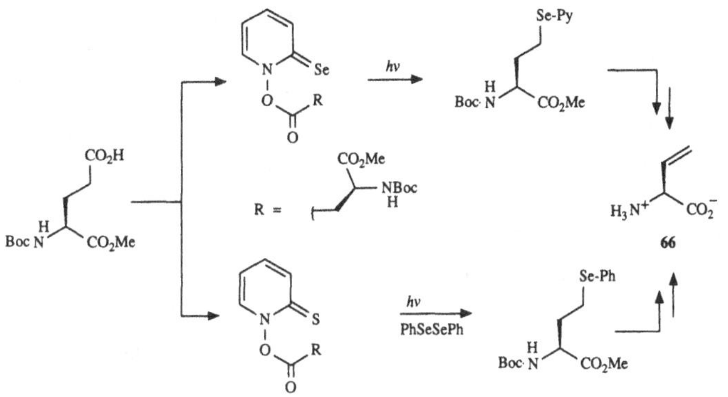

Se-Py
hν
CO_2H
CO_2Me
NBoc
H
R =
Boc N H CO_2Me
Se
O R
O
Boc·N H CO_2Me
H_3N^+ CO_2^-
66
Se-Ph
hν
PhSeSePh
S
Boc·N H CO_2Me

Scheme 26

this method was demonstrated when the presence of alcoholic, phenolic and even indolic groups could be tolerated in this reaction. Manipulation of side chain carboxyl groups was also accomplished after appropriate protection of the α-carboxyl functionality. As shown in Scheme 26, synthesis of an important amino acid, vinyl glycine **66**, was accomplished in optically pure form from readily available glutamic acid.[72] Interception of carbon radicals by suitably activated olefins provides an efficient process for increasing the chain length. This was exemplified in highly desirable transformations like the ones shown in Scheme 27. The esters derived from the

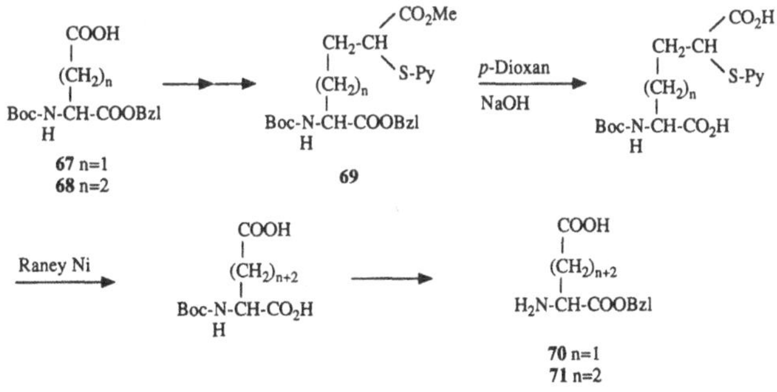

COOH
$(CH_2)_n$
Boc-N-CH-COOBzl
H
67 n=1
68 n=2
CO_2Me
CH_2-CH
S-Py
$(CH_2)_n$
Boc-N-CH-COOBzl
H
69
p-Dioxan
NaOH
CO_2H
CH_2-CH
S-Py
$(CH_2)_n$
Boc-N-CH-CO_2H
H
Raney Ni
COOH
$(CH_2)_{n+2}$
Boc-N-CH-CO_2H
H
COOH
$(CH_2)_{n+2}$
H_2N-CH-COOBzl
70 n=1
71 n=2

Scheme 27

protected amino acids **67** and **68** are photolysed in the presence of methyl acrylate. The resulting addition products **69** are subjected to successive steps of saponification, reduction and deprotection to give the optically pure α-aminoadipic **70** and α-aminopimelic **71** acids.[73] Syntheses of the two most important seleno-amino acids, L-selenomethionine **72** and L-seleno-cysteine **73** have been carried out using this methodology starting with the readily available glutamic and aspartic acid derivatives. Irradiation of the acyl derivative in the presence of excess dimethyl diselenide afforded the selenomethionine derivative which was deprotected using the conditions shown in Scheme 28. An alternative route to compound **72** involved preparation of bromo derivative **74** and displacement of the bromide using sodium methyl selenide.[37] In the synthesis of selenocysteine, Scheme 29, dicyanogen triselenide was found to

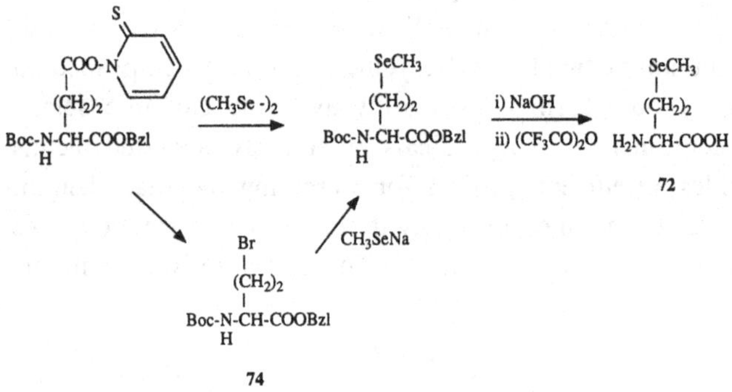

S
COO-N
$(CH_2)_2$
Boc-N-CH-COOBzl
H

$(CH_3Se\text{-})_2$ →

$SeCH_3$
$(CH_2)_2$
Boc-N-CH-COOBzl
H

i) NaOH
ii) $(CF_3CO)_2O$ →

$SeCH_3$
$(CH_2)_2$
H_2N-CH-COOH

72

Br
$(CH_2)_2$
Boc-N-CH-COOBzl
H

74

CH_3SeNa

Scheme 28

S
COO-N
CH_2
Boc-N-CH-COOBzl
H

75

$Se(SeCN)_2$ →

SeCN
CH_2
Boc-N-CH-COOBzl
H

76

i) $NaBH_4$
ii) O_2 →

$Se\text{-})_2$
CH_2
H_2N-CH-COOH

73

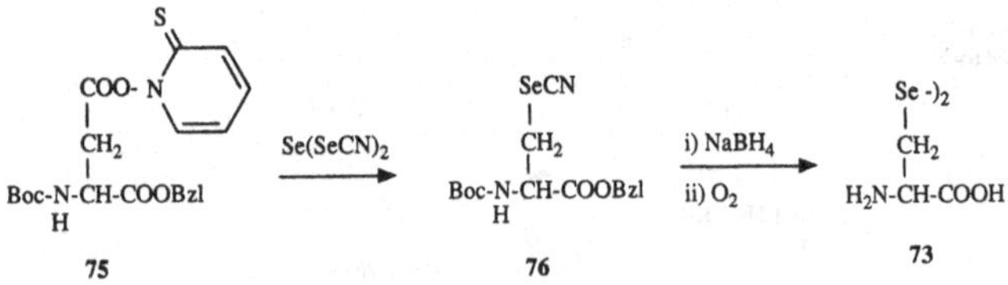

Scheme 29

be very convenient for trapping the radical generated by the photolysis of **75**. The resulting selenocyanate **76** was converted into selenocysteine **73** by treatment with borohydride followed by deprotection.[37]

Previous attempts at the synthesis of perhydroindole-2-carboxylic acid derivatives **77** shown in Scheme 30 have not

{ **78** X = -SPy, R = -Ac, R′ = -^{t}Bu ⟶ **77** X = R = R′ = H }

Scheme 30

been very high yielding. The Barton decarboxylation method permitted an easy access to their synthesis. Straightforward modification of the aspartic acid side chain, Scheme 27, allowed incorporation of the chiral centers of L-aspartic acid into the 2-position of the desired compound with complete retention of the stereocenters.[52] Photolysis of the *N*-hydroxy-2-thiopyridone derivative from **78c** gave the cyclized product as a mixture of two stereoisomers. The stereochemistry assigned to **77b–c** is based on the X-ray study of an analog.[52] Reductive removal of the thiopyridyl function and deprotection using standard procedures gave the desired perhydroindole derivatives **77**.[52]

Until recently, stereocontrolled radical reactions had not been properly investigated. This field has witnessed a rapid growth during the past few years.[74] We (in collaboration with

Dr Géro and Dr Quiclet-Sire) have also looked at stereoselective radical reactions. It was noted from previous experience that stereospecificity can only be accomplished by using one chiral center to direct the formation of another. It is certainly important to note that stereoselective reactions are better controlled in the neighboring positions in a five-membered ring rather than in a six-membered ring.[74] So, with these considerations in mind we, in collaboration with Dr Géro, decided to employ the known *R*,*R*-monoester ketal **79**, readily available from the French National Acid (a compound with a venerable history!).[75] Irradiation of the acyl thiohydroxamate **80** gave the sulfide **81** as the only isomer detectable by NMR. A more rigorous proof of the retention of configuration was secured by studying the addition of the tartaric acid derived radical to methyl acrylate.[76] This gave the adduct **82**. Peracid treatment afforded the sulfoxide, which was heated to 110 °C in refluxing toluene to give the *trans*- olefin **83** after clean thermolysis. Cleavage of the double bond with RuO_2–$NaIO_4$ in acetone–water and methylation with diazomethane gave the dimethyl tartrate derivative **84** identical to an authentic sample of the starting material. The striking conservation of stereochemistry was confirmed by carefully analysing the product mixtures using multidimensional HPLC, which showed a 25:1 ratio of the isomers, good enough for synthetic purposes. An identical sequence of reactions was carried out on the *meso*-derivative **85**. Half hydrolysis of **85** gave monomethyl ester **86**. Addition of the derived radical to methyl acrylate followed by a degradation procedure gave racemic dimethyl tartrate.[76] The stereoselectivity was high enough for most practical purposes, but could undoubtedly be improved, if necessary. This could be accomplished by simply replacing either the methyl ester with a more bulky ester, or by making the ketal function more bulky.

This approach has also been examined with various other olefins. Phenyl vinyl sulfone and *N*-methylmaleimide are nonpolymerizable and thus easier to work with than methyl acrylate (which always gave the background compound **81** and a small percentage of the two-fold adduct). The former gave 70% of **87**, while the latter gave 93% of **88**. Both were, of course, mixtures of stereoisomers since the asymmetric centers created beyond the tartaric acid moiety cannot be controlled.[76] Heating **88** with copper powder resulted in the smooth elimination of pyridyl sulfide group to give the olefin **89** as a single isomer.

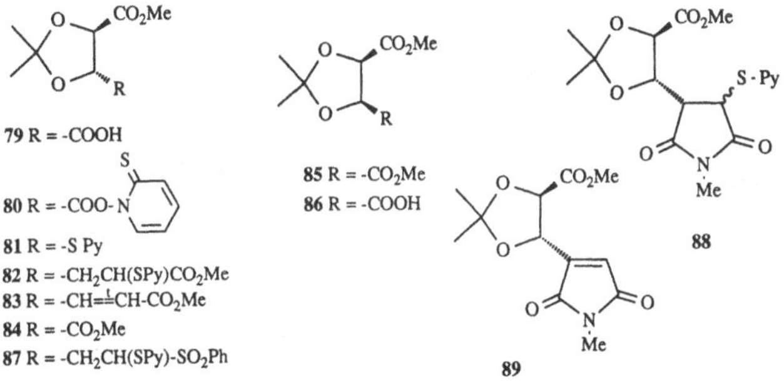

79 R = -COOH
80 R = -COO-N (2-thiopyridone)
81 R = -S Py
82 R = $-CH_2CH(SPy)CO_2Me$
83 R = $-CH{=}CH-CO_2Me$
84 R = $-CO_2Me$
87 R = $-CH_2CH(SPy)-SO_2Ph$

85 R = $-CO_2Me$
86 R = -COOH

88

89

Many important natural products are (formerly) derived by chain elongation at position 5′ of pentoses, or at position 6′ of hexoses. Uronic acids, which are easily prepared, can be converted into the 4′ radical **90** by chemistry based on the thiohydroxamate **6**.[77] We postulated that, if the hindrance on the α-side of the molecule was great enough, the carbon–carbon bond formed by reaction of **90** with a suitable radicophilic olefin would be the natural β-bond. In fact, even a dimethylketal as in **90** (B = natural base or protected derivative thereof) was sufficient to direct the bond formation very largely to the desired β-face.[77] The diacetone ketal of glucuronic acid **91** upon conversion to its *N*-hydroxy-2-thiopyridone derivative **92** and then photolysis in the usual way in the presence of methyl acrylate gave the expected derivative **93** as a mixture of

90

91 X = -COOH

92 X = -COO-N (pyridine-2-thione)

93 X = $-CH_2CH(SPy)CO_2Me$

94 X = $-CH\overset{t}{=}CH-CO_2Me$

95 X = $-CH_2CH(SPy)-SO_2Ph$

96 X = $-CH\overset{t}{=}CH-SO_2Ph$

97 X = -COOH

98 X = $-CH_2CH(SPy)-SO_2Ph$

99 X = $-CH\overset{t}{=}CH-SO_2Ph$

100

101 X = $-CH_2CH(SPy)SO_2Ph$

102 X = $-CH\overset{t}{=}CH-SO_2Ph$

103 X = -COOH

104 X = $-CH_2CH(SPy)SO_2Ph$

105 X = $-CH\overset{t}{=}CH-SO_2Ph$

106 X = -COOH

107 X = $-CH_2CH(SPy)SO_2Ph$

108 X = $-CH\overset{t}{=}CH-SO_2Ph$

diastereomers. Oxidation to sulfoxide and elimination afforded the unsaturated ester **94** as a single compound. Oxidation with RuO_4 gave back pure starting material **91**. A similar series of reactions was carried out using phenyl vinyl sulfone as a radical trap. This afforded the mixed isomers **95** and, after elimination, the pure olefin **96**. Similarly, the ribofuranuronic acid derivative **97** using phenyl vinyl sulfone was converted to a mixture of stereoisomers **98** which on oxidation and elimination gave a single compound **99**. In contrast the D-lyxofuranuronic acid derivative **100** gave, on addition of the radical to phenyl vinyl sulfone, the adducts **101** which on oxidation and elimination afforded a single unsaturated adduct **102** in which the side chain underwent inversion of configuration; this, in turn, provided the enantiomer of **99**. The uridine derivative **103** was a particularly important case. The derived radical gave a good yield of adduct **104** with phenyl vinyl sulfone which, on oxidation and elimination, afforded the vinyl sulfone **105** as a single compound. The adenine derivative **106** afforded the adduct **107** upon photolysis in the presence of phenyl vinyl sulfone. An oxidation and elimination sequence gave the compound **108**. The configuration was ascertained by spectroscopy and confirmed by degradation to the starting acid. The same efficient stereoselectivity was also reported for the reaction of the 4′ radical with diethylvinylphosphonate. Reductive removal of the thiopyridyl group from the adduct and removal of the protecting groups afforded the phosphonates which are isosteric with the corresponding naturally occurring 5′-monophosphates. Also use of the bulky tert-butyldiphenylsilyl group, attached to the 3′-hydroxyl of the uronic acid from thymidine, permitted a stereoselective formation of a carbon–carbon bond using the 4′-carbon radical.

Sinefungin **109** is an important antibiotic[78] with anti-fungal, anti-parasite and strong anti-AIDS activity. It also shows

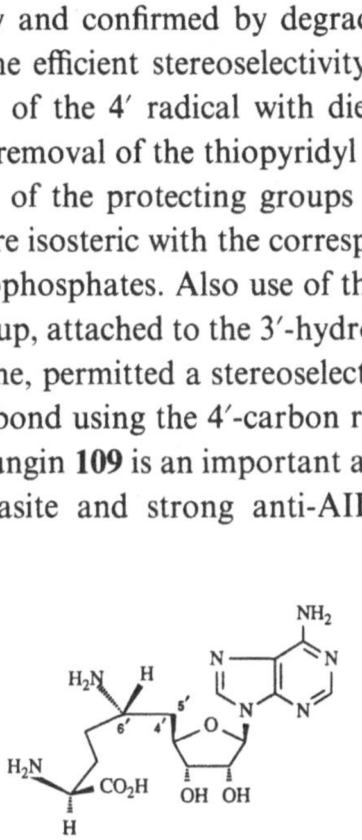

109 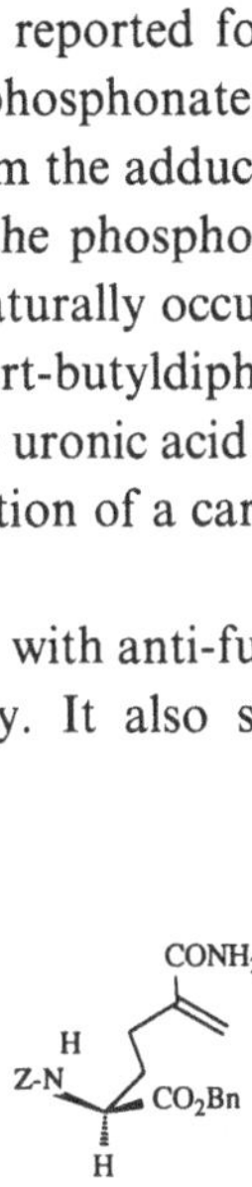 **110**

mammalian toxicity. Until recently, this biological activity could not be evaluated properly through lack of the natural product. We decided[79] to synthesize sinefungin by radical chemistry involving the adenosine derivative **106** and an unsaturated amide **110** readily available from aspartic acid, again using radical chemistry based on **6** (conventional peptide nomenclature is used: Z = carbobenzyloxy, Bn = benzyl). Using the appropriate derivative of **106**, the entire carbon skeleton was constructed in one step by the addition of the 4′-radical to **110**. Known chemistry converted the amide stereospecifically to amine. Removal of the protecting groups then gave the desired sinefungin as well as its epimer at 6′. The biological activity of sinefungin was then studied in detail. Similar studies were also carried out on the uracil analog as well, which was prepared in the same way starting with uridine.[80,81] Another ionic-based synthesis of sinefungin has recently been reported by Rapoport.[82]

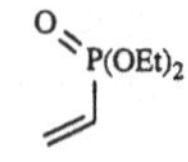

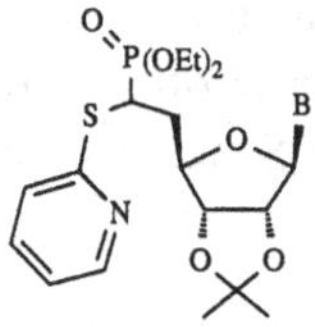

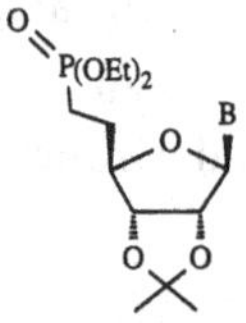

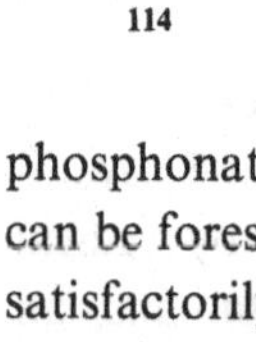

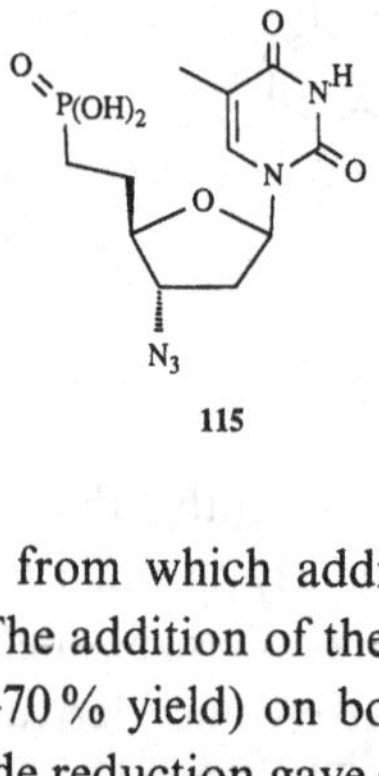

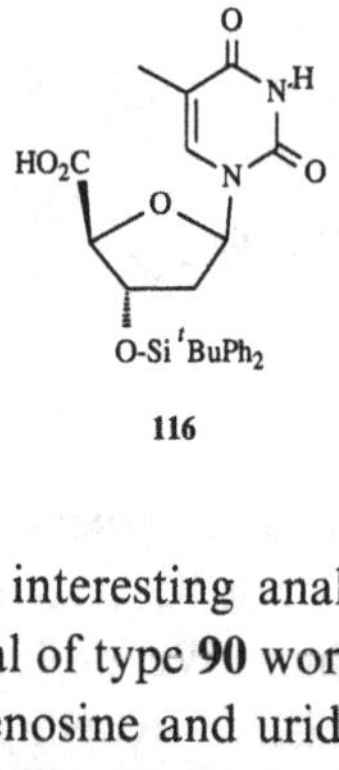

Phosphonates which are isosteric with RNA and DNA derivatives are potentially of great biological interest. It seemed to us[83] that the addition of the radical **90** to diethylvinylphosphonate **111** would afford **112**, easily reducible to **113**, or by oxidation and elimination converted to the vinylphosphonate **114** from which additional interesting analogs can be foreseen. The addition of the radical of typc **90** worked satisfactorily (45–70 % yield) on both adenosine and uridine. Tributyl tin hydride reduction gave cleanly **113** (70–95 %). The reaction could also be applied to aspartic and glutamic acids to give optically active phosphonate derivatives of known biological activity. We decided to make the phosphonate analog **115** of AZT in the hope that it would be a powerful anti-AIDS compound. We employed the uronic acid **116** using tert-butyldiphenylsilyl as a bulky protecting group to direct the radical reaction to the *β*-face of the molecule. This worked well in practice. The phosphonate addition (70%) and further manipulation using known ionic chemistry afforded the desired phosphonic acid **115**.[83] The tests conducted in France, however, did not show anti-AIDS activity! In either event, the work expanded the radical chemistry of thiohydroxamic acid and showed excellent stereoselectivity.

The marked stereoselectivity of these radical reactions must be ascribed to the effect of the bulk of the protecting groups. However, the radicals that were manipulated may have also shown an anomeric effect due to the vicinal oxygen. In a furanose sugar it is not easy to evaluate the anomeric effect without ESR, and also, the molecules concerned have two fused five-membered rings. This fixes the conformation into a V-shape. Such molecules are well known to have *exo*-reactivity and the formation of *endo*-bonds is difficult. All these effects may be acting together to make this nucleoside chemistry especially stereospecific.

The elements with a lone pair of electrons and very rich redox chemistry, have certainly been well exploited in ionic reactions. We decided to study the radical chain reactions based on acyl derivatives of thiohydroxamates and incorporating compounds of group V or VI elements in a unique fashion. In these very highly disciplined and efficient chain reactions (if properly designed and administered) the basic concept is one of valence shell expansion.[11] The general philosophy of this series of reactions developed by us is depicted in Scheme 31. A simple illustration of this concept is the use of sulfur dioxide as a radical trap in conjunction with an *o*-acyl thiohydroxamate leading to the formation of *S*-pyridyl alkylthiosulfonates **117**.[84] For practical reasons this reaction is carried out at −10 °C in a mixture of dichloromethane and liquid SO_2. The sulfur atom undergoes a valence shell expansion from S(IV) to S(VI).

On similar grounds, P(III) to P(V) chemistry was also

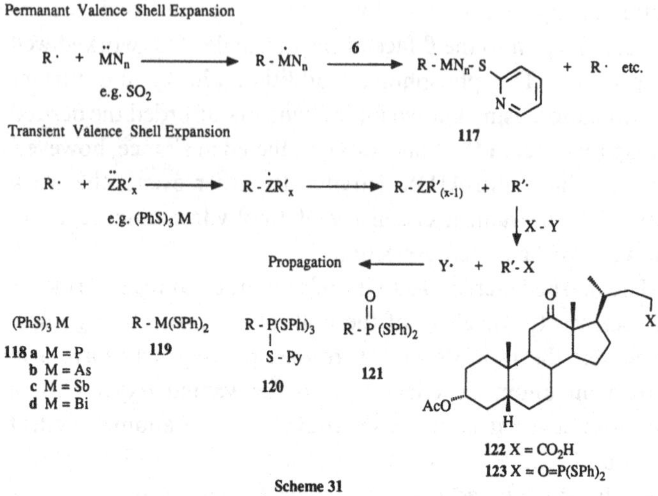

Permanant Valence Shell Expansion

R· + $\ddot{M}N_n$ ⟶ R - $\dot{M}N_n$ —6⟶ R - MN_n- S-(2-pyridyl) + R· etc.

e.g. SO_2

117

Transient Valence Shell Expansion

R· + $\ddot{Z}R'_x$ ⟶ R - $\dot{Z}R'_x$ ⟶ R - $ZR'_{(x-1)}$ + R'·

e.g. $(PhS)_3$ M

X - Y

Propagation ⟵ Y· + R'- X

$(PhS)_3$ M

118a M = P
b M = As
c M = Sb
d M = Bi

R - $M(SPh)_2$
119

R - $P(SPh)_3$ (S - Py)
120

R - P(=O) $(SPh)_2$
121

AcO, H, O, X

122 X = CO_2H
123 X = O=$P(SPh)_2$

Scheme 31

explored. Phosphonic acid analogs of carboxylic acids are readily prepared by reaction of *o*-acyl thiohydroxamates with tris(phenylthio)phosphorus **118a**.[85] The initial radical addition product, alkyl*bis*(phenylthio)-phosphine **119a** undergoes addition of the disulfide byproduct to give a pentavalent phosphorus species **120**, which is hydrolyzed to dithiophosphonate **121** on workup. This sequence was successfully illustrated in the formation of the bile acid analog **123** from 12-ketolithocholic acid ester **122**.[85]

Alkyl radicals also react with group Va trisphenylsulfides to give intermediates of general formula R–$M(SPh)_2$, **119**.[85,86] These react spontaneously with air to give the corresponding alcohols. This procedure is especially useful in the case of antimony. It is sufficient to stir the acyl derivatives of thiohydroxamate with **118c** under air to obtain the nor-alcohol directly and in high yield, Scheme 32. The intermediate organometalloid could also be oxidized with nitrogen dioxide to give the expected nitro-alkane (in modest yield).[86] The corresponding organobismuth intermediate derived from 3,3-

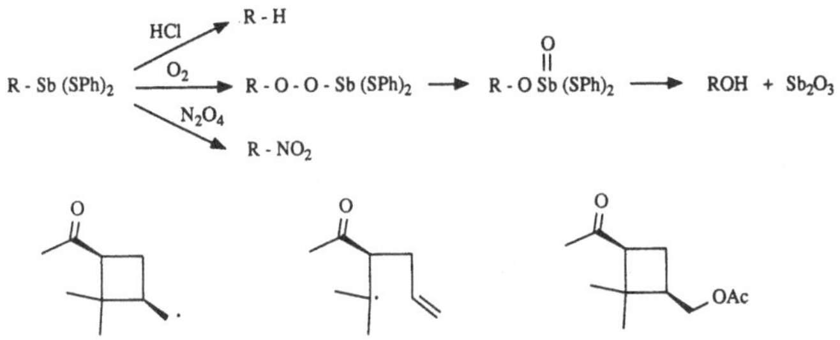

$R\text{-}Sb(SPh)_2 \xrightarrow{HCl} R\text{-}H$

$R\text{-}Sb(SPh)_2 \xrightarrow{O_2} R\text{-}O\text{-}O\text{-}Sb(SPh)_2 \longrightarrow R\text{-}O\overset{O}{\overset{\|}{Sb}}(SPh)_2 \longrightarrow ROH + Sb_2O_3$

$R\text{-}Sb(SPh)_2 \xrightarrow{N_2O_4} R\text{-}NO_2$

OAc

125 **126** **124**

Scheme 32

diphenylpropionic acid could actually be isolated, which provided strong evidence for the proposed mechanism. A simple adaptation of this sequence was illustrated in an easy transformation of *cis*-pinonic acid into the acetate **124**. This compound is an important precursor for the synthesis of **24**, the sex pheromone of the citrus mealybug, as mentioned earlier. The problem with the earlier work in this case was shown to be due to the opening of the radical **125** to give the more stable radical **126** with relief of ring strain.[38] These reactions seem to follow a general pathway: radical addition, transient valence shell expansion, and expulsion of the same or more stable radical. This concept led to the design of radical accumulators.

Radical accumulators whose presence might facilitate addition to *β*-mono and *β*,*β*-disubstituted olefins were conceived. It seemed to us that alkylaryl or dialkyl tellurides should react with alkyl radicals and give an intermediate radical of type $R^1R^2R^3Te^{\bullet}$ (an expanded valence shell) which might have a relatively longer life on the radical time scale. A secondary objective would be the exchange of one radical against another. In this way, the special nucleophilic properties of the aryl telluride anion for example, could be exploited to make complex natural product derived radicals.

The photolysis of **6** ($R = CHMe_2$) in the presence of diisopropyl telluride **127** gave the postulated radical **128**, whose

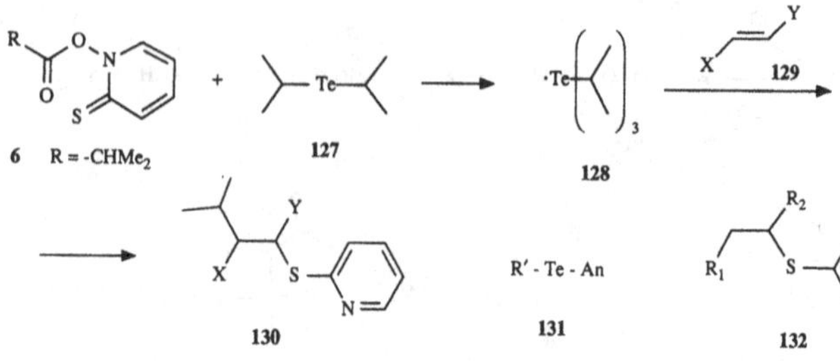

6 R = -CHMe$_2$ 127 128 129 130 R' - Te - An 131 132

Scheme 33

interaction with activated olefins **129** was studied[87] (Scheme 33). With phenyl vinyl sulfone **129** (X = H. Y = SO_2Ph), the adduct **130** (X = H, Y = SO_2Ph) was formed in good yield. However in comparable experiments with **6** (R = $CHMe_2$ and other radicals) without **127**, there was no significant change in yield. Moreover, when a primary radical was generated from **6** (R = Me, $PhCH_2CH_2$, etc.) in the presence of **127**, a clean radical exchange occurred to give $MeTeCHMe_2$ or $PhCH_2CH_2TeCHMe_2$ and adduct **130** (X = H, Y = $PhSO_2$) in satisfactory yield.

So, the exchange process does exist, but there is no observable accumulator effect. The exchange process is, however, useful in the preparation of carbon radicals from complex natural products like carbohydrates.[88] Since dianisyl ditelluride is easy to prepare, we have used the derived ($NaBH_4$) anisyl telluride anion as a nucleophile at primary and secondary positions, including especially the glycosidic carbon, to displace tosylates or bromides to give the appropriate anisyl tellurides, **131**. Photolysis of **6** (R = Me, a 'trigger' reaction) affords a controlled supply of methyl radicals which exchange with tellurides **131** to give AnTeMe (An = anisyl) and the desired carbohydrate radical R′•. In the presence of a suitable radical trap like **129** (X = H, Y = $PhSO_2$, COMe, CO_2Me etc.) adducts **132** (R_1 = carbohydrate residue, R_2 = $PhSO_2$, COMe.

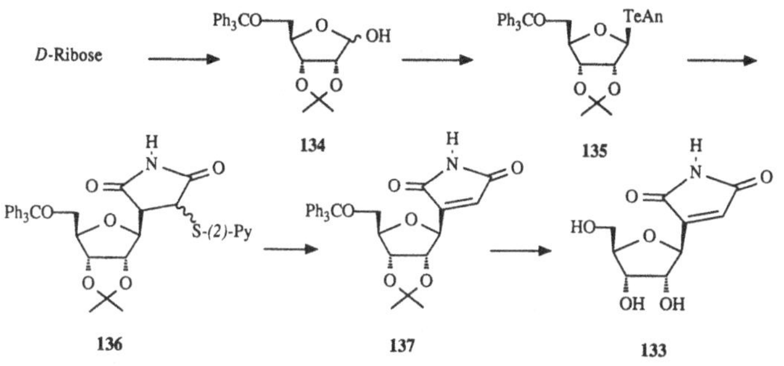

Scheme 34

CO_2Me, etc.) were formed in good yield. A short and high-yielding synthesis of showdomycin **133** (Scheme 34) illustrated the utility of the method.[88] D-Ribose was converted to the known derivative **134** which, on mesylation and displacement with anisyltelluride anion, gave **135**. Methyl radical exchange on **135** in the presence of maleimide gave **136** which, on oxidation to sulfoxide and elimination, afforded the derivative **137**. Standard deprotection procedures readily transformed **137** into the antibiotic showdomycin **133**. The overall yield was about 30 %.

With so much good radical chemistry based on acyl derivatives of thiohydroxamic acids, we naturally wondered if the corresponding derivatives of ordinary hydroxamic acids would show similar reactivity. Of the hydroxamic acids studied, only the dihydrocinnamoyl derivative of **138** gave a good yield (97 %) of the hydrocarbon in a tin hydride-initiated reduction. The next best was the derivative of *N*-hydroxy-2-pyridone which gave 73 % of hydrocarbon. These results,[89] as well as

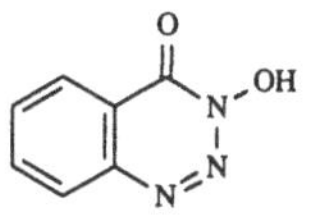

138 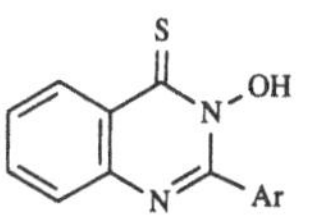 **139**

those in the literature,[90] serve to confirm the superiority of thiohydroxamic acids as radical generators. We were not able to trap any radical produced from **138** to make a carbon–carbon bond.

The reactions based on *N*-hydroxy-2-thiopyridone derivatives are clearly radical chain reactions. In a recent paper,[91] we have reported quantum yield measurements for a number of reactions based on *N*-hydroxy-2-thiopyridone. Most of the reactions had quantum yields of 10–30. Synthesis of the *N*-hydroxyquinazolin-4-thione **139** (Ar = Ph, An, 1-Naph) by an improved route gave thiohydroxamic acids which were more sensitive to light than *N*-hydroxy-2-thiopyridone. The quantum yield for bromination was in the range 30–60. More important, while the *N*-hydroxy-2-thiopyridone system at −30 °C, makes[21] radicals only in a non-chain fashion, the derivatives of **139** continue radical chain reactions even at −60 °C.

Although these synthetic transformations have been possible using the radical chemistry of the thiocarbonyl function, we still feel that a substantial harvest of important reactions remains to be garnered.

References

1. a) Barton, D. H. R. *Aldrichimica Acta* (1990), **23**, 3. b) Barton, D. H. R. & McCombie, S. W. *J. Chem. Soc., Perkin Trans. I* (1975), 1574.
2. a) Wiberg, K. B., Lowry, B. R. & Colby, T. H. *J. Am. Chem. Soc.* (1961), **83**, 3998. b) Eaton, P. E. & Cole, T. W. *J. Am. Chem. Soc.* (1964), **86**, 3157.
3. a) Langhals, H. & Rüchardt, C. *Chem. Ber.* (1975), **108**, 2156. b) Pomerantz, M. & Dassanayake, N. L. *J. Am. Chem. Soc.* (1980), **102**, 678.
4. a) Barton, D. H. R., George, M. V. & Tomoeda, M. *J. Chem. Soc.* (1962), 1967. b) Billingham, N. C., Jackson, R. A. & Malek, F. *J. Chem. Soc., Perkin Trans. I* (1979), 1137.
5. a) Della, E. W. & Patney, H. K. *Synthesis* (1976), 251. b) Della,

E. W. & Patney, H. K. *Aust. J. Chem.* (1976), **29**, 2469.

6. a) Barrett, A. G. M., Barton, D. H. R., Bielski, R. & McCombie, S. W. *J. Chem. Soc., Chem. Commun.* (1977), 866. b) Barrett, A. G. M., Barton, D. H. R., Bielski, R. & McCombie, S. W. *J. Chem. Soc., Perkin Trans. I* (1979), 2378. c) Lythgoe, B. & Waterhouse, I. *Tetrahedron Lett.* (1977), 4223. d) Boothe, T. E., Green, J. L. & Shevlin, P. B. *J. Org. Chem.* (1980), **45**, 794. e) Julia, S. & Lorne, R. C. R. *Hebd. Seances Acad. Sc. Ser. C* (1971), **273**, 174.
7. Barton, D. H. R., Dowlatshahi, H. A., Motherwell, W. B. & Villemin, D. *J. Chem. Soc., Chem. Commun.* (1980), 732.
8. a) Walter, W. & Schaumann, E. *Synthesis* (1971), 111. b) Sandler, S. R. & Kato, W. *Org. Funct. Group Prepn.* Academic Press: New York, 1972; Vol **3**, p. 433.
9. Barton, D. H. R. In *Some Recollections of Gap Jumping*, part of the series: *Profiles, Pathways and Dreams* Ed. Seeman, J. I.; American Chemical Society, Washington, D.C., 1991.
10. Crich, D. *Aldrichimica Acta* (1987), **20**, 35.
11. Crich, D. & Quintero, L. *Chem. Rev.* (1989), **89**, 1413.
12. Barton, D. H. R., Crich, D. & Motherwell, W. B. *J. Chem. Soc., Chem. Commun.* (1983), 939.
13. Barton, D. H. R. & Zard, S. Z. *Phil. Trans. R. Soc. Lond. B* (1985), **311**, 505.
14. Barton, D. H. R. & Motherwell, W. B. *Heterocycles* (1984), **21**, 1.
15. Barton, D. H. R., Crich, D. & Motherwell, W. B. *Tetrahedron* (1985), **41**, 3901.
16. Ingold, K. U., Lusztyk, J., Maillard, B. & Walton, J. C. *Tetrahedron Lett.* (1988), **29**, 917.
17. Barton, D. H. R., Hervé, Y., Potier, P. & Thierry, J. *Tetrahedron* (1988), **44**, 5479.
18. Tenud, L., Farooq, S., Seibl, J. & Eschenmoser, A. *Helv. Chim. Acta* (1970), **53**, 2059.
19. Ankers, W. B., Brown, C., Hudson, R. F. & Lawson, A. J. *J. Chem. Soc., Chem. Commun.* (1972), 935.
20. Barton, D. H. R., Crich, D. & Potier, P. *Tetrahedron Lett.* (1985), 5943.

21. Barton, D. H. R., Bridon, D., Fernandez-Picot, I. & Zard, S. Z. *Tetrahedron* (1987), **43**, 2733.
22. a) Ingold, K. U. & Beckwith, A. L. J. In *Rearrangements in Ground and Excited States*; De Mayo, P., Ed.; Academic Press: New York, 1980 and references cited therein. b) Beckwith, A. L. J. *Tetrahedron* (1981), **37**, 3073. c) Beckwith, A. L. J., Easton, J. C., Lawrence, T. & Serelis, A. K. *Aust. J. Chem.* (1983), **36**, 545. d) Beckwith, A. L. J. & Hay, B. P. *J. Am. Chem. Soc.* (1989), **111**, 230. e) Lusztyk, J., Maillard, B., Deycard, S., Lindsay, D. A. & Ingold, K. U. *J. Org. Chem.* (1987), **52**, 3509. f) Newcomb, M. & Park, S.-U. *J. Am. Chem. Soc.* (1986), **108**, 4132. g) Newcomb, M. & Kaplan, J. *Tetrahedron Lett.* (1987), **28**, 1615. h) Newcomb, M. & Kaplan, J. *Tetrahedron Lett.* (1988), **29**, 3449. i) Newcomb, M. & Glenn, A. G. *J. Am. Chem. Soc.* (1989), **111**, 275. j) Newcomb, M. & Manek, M. B. *J. Am. Chem. Soc.* (1990), **112**, 9662. k) Park, S.-U., Varick, T. R. & Newcomb, M. *Tetrahedron Lett.* (1990), **31**, 2975. l) Newcomb, M., Manek, M. B. & Glenn, A. G. *J. Am. Chem. Soc.* (1991), **113**, 949. m) Curran, D. P., Bosch, E., Kaplan, J. & Newcomb, M. *J. Org. Chem.* (1989), **54**, 1826. n) Bohne, C., Boch, R. & Scaiano, J. C. *J. Org. Chem.* (1990), **55**, 5414.
23. Barton, D. H. R., Faro, H. P., Serebryakov, E. P. & Woolsey, N. F. *J. Chem. Soc.* (1965), 2438.
24. a) Kochi, J. K. *J. Am. Chem. Soc.* (1965), **87**, 2500. b) Cristol, J. S. & Firth, W. C. *J. Org. Chem.* (1961), **26**, 280. c) McKillop, A., Bromley, D. & Taylor, E. C. *J. Org. Chem.* (1969), **34**, 1172.
25. For a review, see Johnson, R. G. & Ingham, R. K. *Chem. Rev.* (1956), **56**, 219.
26. Barton, D. H. R., Crich, D. & Motherwell, W. B. *Tetrahedron Lett.* (1983), 4979.
27. Barton, D. H. R. & Zard, S. Z. *Phil. Trans. R. Soc. Lond.*, (1985), **311B**, 505.
28. Barton, D. H. R. & Zard, S. Z. *Pure Appl. Chem.* (1986), **58**, 675.
29. Barton, D. H. R. & Ozbalik, N. *Phosphorous Sulfur Silica* (1989), **43**, 349.
30. Ramaiah, M. *Tetrahedron* (1987), **43**, 3541.
31. Curran, D. P. *Synthesis* (1988), 417 and 489.
32. Private communication.

33. a) Barton, D. H. R., Lacher, B. & Zard, S. Z. *Tetrahedron Lett.* (1985), **26**, 5939. b) Private communications, and see also Vogel, E., Schieb, T., Schultz, W. H., Schmidt, K., Schmicklen, H. & Lex, J. *Angew. Chem., Int. Ed. Engl.* (1986), **25**, 723.
34. Vogel, E., Schieb, T., Schultz, W. H., Schmidt, K., Schmicklen, H. & Lex, J. *Tetrahedron* (1987), **43**, 4321.
35. Barton, D. H. R., Bridon, D. & Zard, S. Z. *Tetrahedron Lett.* (1984), 5777.
36. Barton, D. H. R., Bridon, D. & Zard, S. Z. *Heterocycles* (1987), **25**, 449.
37. Barton, D. H. R., Bridon, D., Hervé, Y., Potier, P., Thierry, J. & Zard, S. Z. *Tetrahedron* (1986), **42**, 4983.
38. Barton, D. H. R., Ozbalik, N. & Schmitt, M. *Tetrahedron Lett.* (1989), **30**, 3263.
39. Barton, D. H. R., Crich, D. & Motherwell, W. B. *J. Chem. Soc., Chem. Commun.* (1984), 242.
40. Hart, D. J. *Science* (1984), **223**, 4639.
41. a) Giese, B. *Angew. Chem. Int. Ed. Engl.* (1983), **22**, 753. b) Giese, B. *Angew. Chem. Int. Ed. Engl.* (1985), **24**, 553.
42. Giese, B. In *Radicals in Organic Synthesis: Formation of Carbon–Carbon Bonds*; Pergamon: Oxford, 1986.
43. Barluenga, J. & Yus, M. *Chem. Rev.* (1988), **88**, 487.
44. Barton, D. H. R. & Crich, D. *Tetrahedron Lett.* (1984), **25**, 2787.
45. Barton, D. H. R. & Crich, D. *J. Chem. Soc., Perkin Trans. I* (1986), 1613.
46. a) Keck, G. E. & Yates, J. B. *J. Am. Chem. Soc.* (1982), **104**, 5829. b) Keck, G. E. & Yates, J. B. *J. Org. Chem.* (1982), **47**, 3590. c) Webb, R. R. & Danishefsky, S. *Tetrahedron Lett.* (1983), **24**, 1357. d) Keck, G. E., Inholm, E. J. & Kochensky, D. F. *Tetrahedron Lett.* (1984), **25**, 1867.
47. Barton, D. H. R., Togo, H. & Zard, S. Z. *Tetrahedron* (1985), **42**, 5507.
48. Barton, D. H. R., Crich, D. & Kretzschmar, G. *Tetrahedron Lett.* (1984), **25**, 1055.
49. Barton, D. H. R., Togo, H. & Zard, S. Z. *Tetrahedron Lett.* (1985), **26**, 6349.
50. Barton, D. H. R., Togo, H. & Zard, S. Z. *J. Chem. Soc., Perkin Trans. I* (1986), 39.

51. Barton, D. H. R., Lacher, B. & Zard, S. Z. *Tetrahedron* (1986), **42**, 2325.
52. Barton, D. H. R., Guilhem, J., Hervé, Y., Potier, P. & Thierry, J. *Tetrahedron Lett.* (1987), **28**, 1413.
53. Barton, D. H. R., da Silva, E. & Zard, S. Z. *J. Chem. Soc., Chem. Commun.* (1988), 285.
54. Barton, D. H. R. & Sarma, J. C. *Tetrahedron Lett.* (1990), **31**, 1965.
55. Castagnino, E., Corsano, S. & Barton, D. H. R. *Tetrahedron Lett.* (1989), **30**, 2883.
56. Barton, D. H. R., Ozbalik, N. & Vacher, B. *Tetrahedron* (1988), **44**, 3501.
57. Shakak, I. & Sasson, Y. *J. Am. Chem. Soc.* (1973), **95**, 3440.
58. Severengiz, T. & Du Mont, W. W. *J. Chem. Soc., Chem. Commun.* (1987), 820; and references cited therein.
59. Cussans, N. J., Ley, S. V. & Barton, D. H. R. *J. Chem. Soc., Perkin Trans. I* (1980), 1650.
60. a) Pews, R. G. & Evans, T. E. *J. Chem. Soc., Chem. Commun.* (1971), 1397. b) Fang, J.-M. & Chen, M.-Y. *Tetrahedron Lett.* (1987), **28**, 2853. c) Fang, J.-M., Chen, M.-Y., Cheng, M.-C, Lee, G.-H., Wang, Y. & Peng, S.-M. *J. Chem. Research (S)* (1989), 272.
61. Barton, D. H. R., Jaszberenyi, J. C. & Theodorakis, E. A. *Tetrahedron Lett.* (1991), **32**, 3321.
62. Barton, D. H. R., Ozbalik, N. & Vacher, B. *Tetrahedron* (1988), **44**, 7385.
63. Barton, D. H. R., Bridon, D. & Zard, S. Z. *Tetrahedron* (1987), **43**, 5307.
64. Barton, D. H. R. & Sas, W. *Tetrahedron* (1990), **46**, 3419.
65. Barton, D. H. R., Chern, C.-Y. & Jaszberenyi, J. C. *Tetrahedron Lett.* (1991), **32**, 3309.
66. Perkow, W., Ullerich, K. and Meyer, F. *Naturwiss.* (1952), **39**, 353.
67. Barton, D. H. R., Boivin, J., Sarma, J. C., da Silva, E. & Zard, S. Z. *Tetrahedron Lett.* (1989), **30**, 4237; and Barton, D. H. R., Boivin, J., Sarma, J. C., da Silva, E. & Zard, S. Z. *Tetrahedron* (1991), **47**, 7091.
68. Barton, D. H. R. & Crich, D. *J. Chem. Soc., Chem. Commun.* (1984), 774.

69. a) Minisci, F., Caronna, T., Cercere, M., Galli, R. & Malatests, V. *Tetrahedron Lett.* (1968), 5609. b) Minisci, F., Zammori, P., Bernardi, R., Cerere, M. & Galli, R. *Tetrahedron* (1970), **26**, 4153. c) Minisci, F. *Synthesis* (1973), 1. d) Minisci, F. *Acc. Chem. Res.* (1975), **8**, 165. e) Minisci, F., Citterio, A. & Giordano, C. *Acc. Chem. Res.* (1983), **16**, 27. f) Citterio, A., Gentile, A. & Minisci. F. *Tetrahedron Lett.* (1982), **23**, 5587. g) Citterio, A., Gentile, A., Minisci, F., Navarrini, V., Sarravalle, M. & Ventura, S. *J. Org. Chem.* (1984), **49**, 4479. h) Castaldi, G., Minisci, F., Tortelli, V. & Vismara, E. *Tetrahedron Lett.* (1984), **25**, 3897. i) Minisci, F., Vismara, E., Morini, G., Fontana, F., Levi, S., Serravalle, M. & Giordano, C. *J. Org. Chem.* (1986), **51**, 474. j) Fontana, F., Minisci, F. & Vismara, E. *Tetrahedron Lett.* (1988), **29**, 1975. k) Fontana, F., Minisci, F. & Vismara, E. *J. Org. Chem.* (1989), **54**, 5224. l) Fontana, F., Minisci, F. & Vismara, E. *J. Heterocycl. Chem.* (1990), **27**, 79 and references cited therein.
70. a) Barton, D. H. R., Garcia, B., Togo, H. & Zard, S. Z. *Tetrahedron Lett.* (1986), **27**, 1327. b) Castagnino, E., Corsano, S., Barton, D. H. R. & Zard, S. Z. *Tetrahedron Lett.* (1986), **27**, 6337.
71. Barton, D. H. R., Hervé, Y., Potier, P. & Thierry, J. *J. Chem. Soc., Chem. Commun.* (1984), 1298.
72. Barton, D. H. R., Crich, D., Hervé, Y., Potier, P. & Thierry, J. *Tetrahedron* (1985), **41**, 4347.
73. a) Barton, D. H. R., Hervé, Y., Potier, P. & Thierry, J. *Tetrahedron* (1987), **43**, 4297. b) Yolande Hervé, Thése de Docteur de l'Université Paris XI, Orsay en Seine, 1986.
74. a) Giese, G. *Angew. Chem. Int. Ed. Engl.* (1989), **28**, 753 and references cited therein. b) Curran, D. P., Shen, W., Zhang, J. & Heffner, T. A. *J. Am. Chem. Soc.* (1990), **112**, 6738. c) Porter, N. A., Swann, E., Nally, J. & McPhail, A. T. *J. Am. Chem. Soc.* (1990), **112**, 6740. d) Giese, B., Zehnder, M., Roth, M. & Zeitz, H.-G. *J. Am. Chem. Soc.* (1990), **112**, 6741. e) Porter, N. A., Scott, D. M. & McPhail, A. T. *Tetrahedron Lett.* (1990), **31**, 1679. f) Rajanbanbu, T. V. *Acc. Chem. Res.* (1991), **24**, 139 and references cited therein.
75. Musich, J. A. & Rapoport, H. *J. Am. Chem. Soc.* (1978), **100**, 48.
76. a) Barton, D. H. R., Gateau-Oleskar, A., Géro, S. D., Lacher, B., Tachdjian, C. & Zard, S. Z. *J. Chem. Soc., Chem. Commun.*

(1987), 1790. b) Porziemsky, J.-P., Krstulovič, A. M., Wick, A., Barton, D. H. R., Tachdjian, C., Gateau-Oleskar, A. & Géro, S. D. *J. Chromatography* (1988), **440**, 183.

77. Barton, D. H. R., Géro, S. D., Quiclet-Sire, B. & Samadi, M. *J. Chem. Soc., Chem. Commun.* (1988), 1372.
78. Hamil, R. L. & Hoehn, M. M. *J. Antibiot.* (1973), **26**, 463.
79. Barton, D. H. R., Géro, S. D., Quiclet-Sire, B. & Samadi, M. *J. Chem. Soc., Perkin Trans. I* (1991), 981.
80. Barton, D. H. R., Géro, S. D., Lawrence, F., Robert-Géro, M., Quiclet-Sire, B. & Samadi, M. *J. Med. Chem.* (1991), in press.
81. Barton, D. H. R., Géro, S. D., Quiclet-Sire, B. & Samadi, M. *J. Chem. Soc., Chem. Commun.* (1989), 1000.
82. Maguire, M. P., Feldman, P. L. & Rapoport, H. *J. Org. Chem.* (1990), **55**, 948.
83. Barton, D. H. R., Géro, S. D., Quiclet-Sire, B. & Samadi, M. *Tetrahedron Lett.* (1989), **30**, 4969.
84. Barton, D. H. R., Lacher, B., Misterkiewicz, B. & Zard, S. Z. *Tetrahedron* (1988), **44**, 1153.
85. Barton, D. H. R., Bridon, D. & Zard, S. Z. *Tetrahedron Lett.* (1986), 4309.
86. Barton, D. H. R., Bridon, D. & Zard, S. Z. *J. Chem. Soc., Chem. Commun.* (1985), 1066.
87. Barton, D. H. R., Ozbalik, N. & Sarma, J. C. *Tetrahedron Lett.* (1988), **29**, 6581.
88. Barton, D. H. R. & Ramesh, M. *J. Am. Chem. Soc.* (1990), **112**, 891.
89. Barton, D. H. R., Blundell, P. & Jaszberenyi, J. C. *Tetrahedron Lett.* (1989), **30**, 2341.
90. a) Hasebe, M., Kogawa, K. & Tsuchiya, T. *Tetrahedron Lett.* (1984), **25**, 3887. b) Hasebe, M. & Tsuchiya, T. *Tetrahedron Lett.* (1986), **27**, 3239. c) Hasebe, M. & Tsuchiya. T. *Tetrahedron Lett.* (1987), **28**, 6207. d) Hasebe, M. & Tsuchiya, T. *Tetrahedron Lett.* (1988), **29**, 6287. e) Okada, K., Okamoto, K. & Oda, M. *J. Am. Chem. Soc.* (1988), **110**, 8736.
91. Barton, D. H. R., Blundell, P. & Jaszberenyi, J. C. *J. Am. Chem. Soc.* (1991). **113**, 6937.

6

Some recent synthetic applications of Barton radical methodologies

SHYAMAL I. PAREKH

Free radical reactions have had a rather rapid and yet far-reaching impact on synthetic planning in organic synthesis in recent years. This is due to the realization of certain obvious advantages that radical methods enjoy over classical ionic reactions. Tandem radical cyclization for the construction of polycyclic ring systems and a large number of applications of this powerful strategy that can be found in the literature illustrates the above point quite adequately. The foundation of such strategic synthetic planning relies heavily on the synthetic tools and the relevant mechanistic understanding available at one's disposal. The chemistry discussed in the last two chapters has dealt with the various radical reactions associated with the thiocarbonyl group. Their invention, the following successful development of various synthetic applications, and the resulting mechanistic understanding of these reactions have undoubtedly made it a very important part of the organic synthetic tools in general. Though this is still a developing field, recently there has been an increasing number of reports on synthetic applications, adaptations or modifications of Barton radical methodologies, which justifies a summary of the same. There are, of course, several excellent treatises available in the literature, where a detailed picture of the reactivity, selectivity, and stability of many types of organic radicals are portrayed.[1] For synthetic utility radical chain reactions are essential, and recently, Giese has provided excellent treatments of various radical methodologies both from a theoretical and synthetic standpoint.[2] With respect to the present discussion, Ramaiah,[3] Curran,[4] Hart,[5] and Crich[6] (to varied lengths) have discussed several examples which portray the multitude of functional groups and skeletal types which can be easily accommodated by the Barton radical methodologies. Hartwig's review provides a very exhaustive discussion on the radical deoxygenation methods.[7a] An in-depth discussion on the syntheses of trideoxyhexoses (and the derivatives thereof) was provided in the review by Hauser and Ellenberger.[7b] Since it would be beyond the scope of this chapter to provide discussion at length on all the published work which has utilized and thus further expanded the scope of Barton radical deoxygenation protocol, the second part of the following discussion merely serves as a summary of the types of applications this has seen and a compilation of the recent relevant references. The first part, however, deals with the synthetic applications of the Barton decarboxylation reaction, and is intended to be demonstrative of the synthetic potential and generality of the radical chemistry associated with thiohydroxamic acid derivatives. The acyl derivatives of *N*-hydroxy-2-thiopyridone **1** have been called Barton esters or PTOC (*p*yridine *t*hione *o*xy-*c*arbonyl) derivatives.

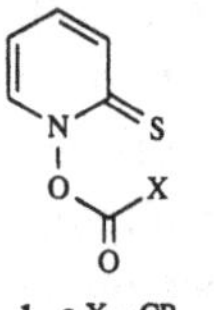

1 a $X = CR_3$
b $X = NR_2$
c $X = OR$

Part I

Radical chemistry as it is practised and perceived today is still largely in terms of tin hydride, distannane or tetra-alkyl stannane mediated reactions, where a trialkyltin radical is the chain transfer agent. This radical has the important ability to generate radicals selectively by abstraction or addition and to transfer chains rapidly by elimination or hydrogen atom donation. A totally different and novel option is provided by the thiohydroxamic acid esters. In this case the chain transfer agent is generated by the fragmentation process and there are distinct advantages to this process which have been discussed earlier. The choice of subsequent functionalization option makes the Barton decarboxylation reaction very useful for a large number of functional group interconversions. The experimental procedure is quite simple, the method is very general, and a wide variety of transformations are possible. As we saw in the previous chapter, the method has been adapted not only for reducing carboxylic acids to nor-alkanes but also for making carbon–carbon or carbon–heteroatom bonds by interception of the intermediate radical.

It is due to this realization that a large spectrum of synthetic applications of Barton decarboxylation can be found in the recent literature. It is important to note that, in most of these synthetic applications and adaptations, the authors have reported good to excellent yields during the decarboxylative radical trapping stage. Moreover, in several papers the researchers report having tried various other methods before trying the Barton decarboxylation approach and concluded that it gave less cumbersome reaction mixtures than the ionic reactions employed for the same purpose.

Pinhey utilized the decarboxylative rearrangement of **2** to the nor-alkyl pyridyl sulfide **3**, oxidation to the corresponding sulfoxide and thermal elimination sequence in a simple transformation of podocarpic acid **4** into useful synthons **5** (for example, in vitamin D_3 synthesis) for steroid CD-ring systems, as shown in Scheme 1.[8]

Reductive decarboxylation has been found to be very useful for the conversion of $R{-}CO_2H \rightarrow R{-}H$ by several research

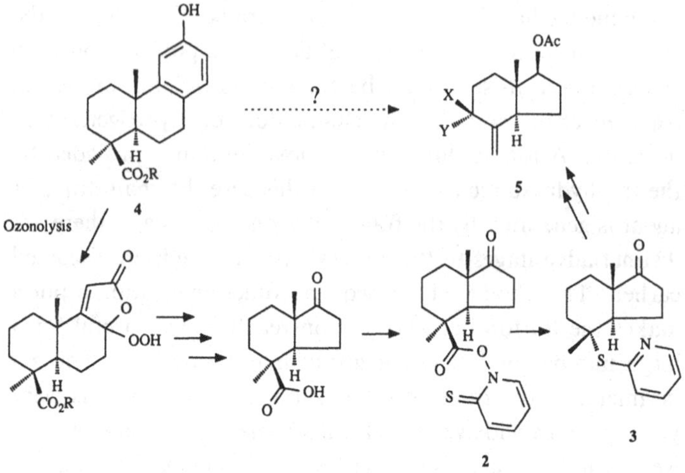

Scheme 1

groups, where a variety of structural features and the presence of different functionalities were easily tolerated. Winkler and coworkers for example, in their investigations on inside–outside intrabridgehead stereoisomerism of the bi- and tricyclic ring systems **6** (established via the intramolecular dioxolenone photocycloaddition), and their applications in the synthesis of taxol and ingenane, have found it quite convenient to reduce the carboxyl group via the Barton protocol during various stages of their investigations as depicted in Schemes 2a–b.[9] In a formal synthesis of vindorosine, **7**, starting from L-tryptophan, the intramolecular vinylogous amide photocycloaddition, retro-Mannich fragmentation, Mannich closure sequences were used, where Barton decarboxylation was found useful in **8a** → **8b** sequence, Scheme 2c.[10] Stereocontrolled transannular radical cyclization by Winkler has provided a new approach to the asymmetric synthesis of linearly fused cyclopentanoids, and again the reductive decarboxylation was successfully utilized for **9a** → **9b** as exemplified in Scheme 2d.[11]

In their syntheses of trishomohypostrophene **10a**, trishomopentaprismane **11a**, and the corresponding ketones **10b** and **11b**

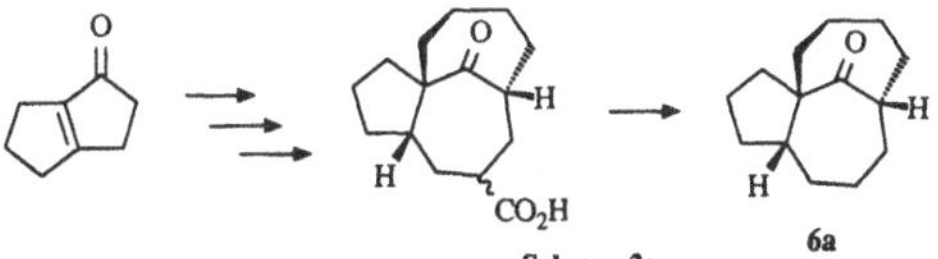

O
H
CO2H
6a

Scheme 2a

R = CO2H → R = H

6b **6c**

Scheme 2b

Ac
Et
OAc
OH
CO2Me
Me

8a R = CO_2H → **8b** R = H **7**

Scheme 2c

9a X or Y = CO2H
9b X or Y = H

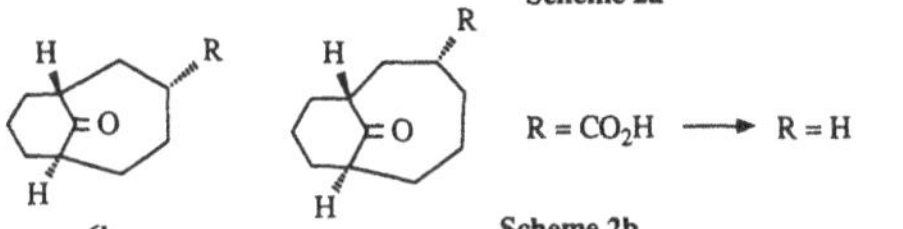

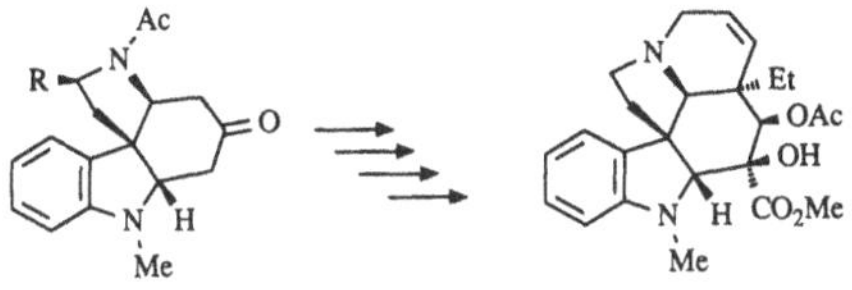

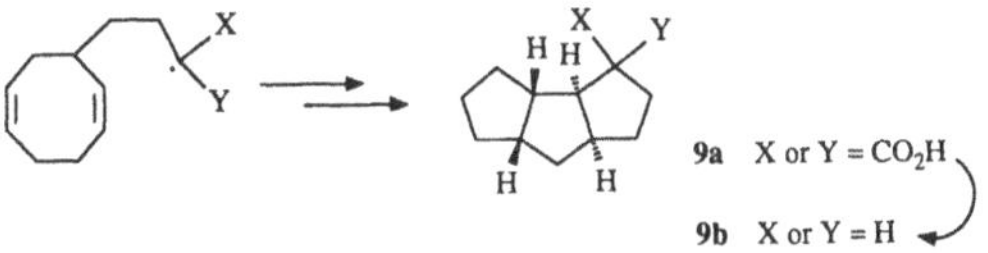

Scheme 2d

(starting with the pentacyclic diester **12**), Musso and Boland favored the Barton reductive decarboxylation reaction as depicted in Scheme 3.[12]

Liu's recently developed methodology for the asymmetric synthesis of 1,2,3,4-tetrahydro-*β*-carbolines **13a**, was extended

OH
12
HO2C
CO2H
X = Y = OCH3
10a X,Y = H
10b X,Y = =O
11a X,Y = H
11b X,Y = =O

Scheme 3

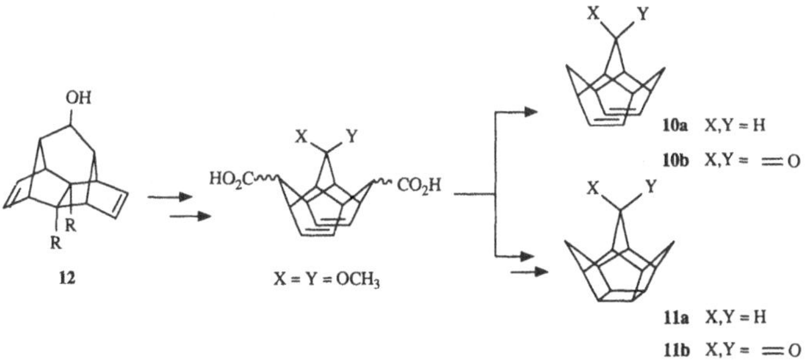

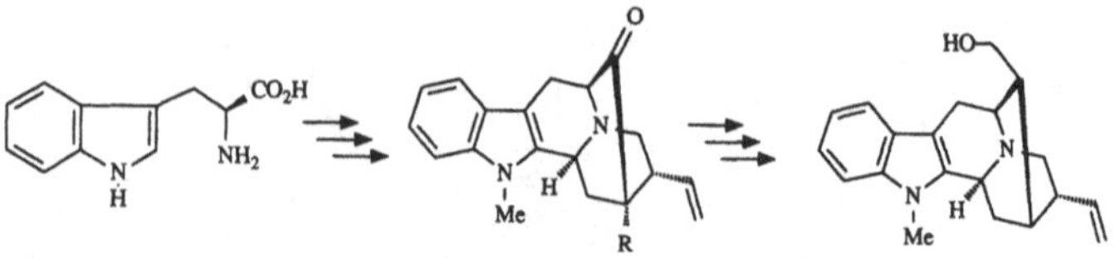

13b R = CO_2H
13a R = H
14

Scheme 4

for the total synthesis of N_a-methyl-Δ^{18}-isokoumidine **14** (a possible precursor for koumine type indole alkaloids) where L-tryptophan was used as the starting material, Scheme 4.[13] Here many methods to remove the bridgehead carboxyl group in **13b** → **13a** sequence by conventional methods had failed, and finally only Barton decarboxylation was found to be successful.

Musso has reported the synthesis of diasterane (tricyclo-[3.1.1.1^{2,4}]octane) **15**. For this first member of the series of asteranes, the decarboxylation of **16b** → **16c** was best achieved via the photolysis of the Barton ester of **16a** in the presence of ^{t}BuSH, as shown in Scheme 5.[14] Fukumoto has accomplished asymmetric total synthesis of atisine **17**, where the bridged pentacyclic intermediate **18**, a precursor for atisine, was synthesized via an intramolecular double Michael reaction starting with **19**, Scheme 6.[15] Barton protocol was favored during the late stages of the synthesis and the presence of various functionalities was easily accommodated.

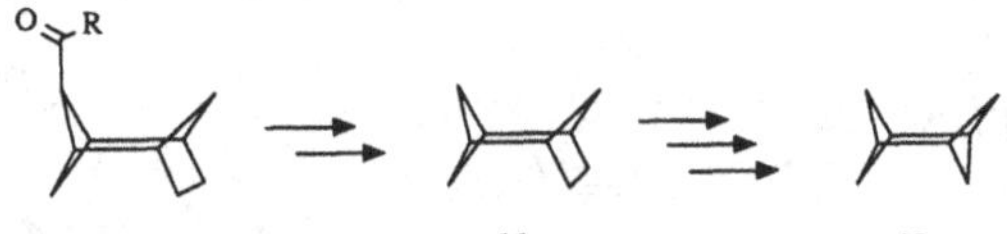

16a R = OH
16b R = O—N
16c
15

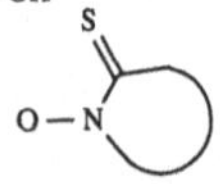

Scheme 5

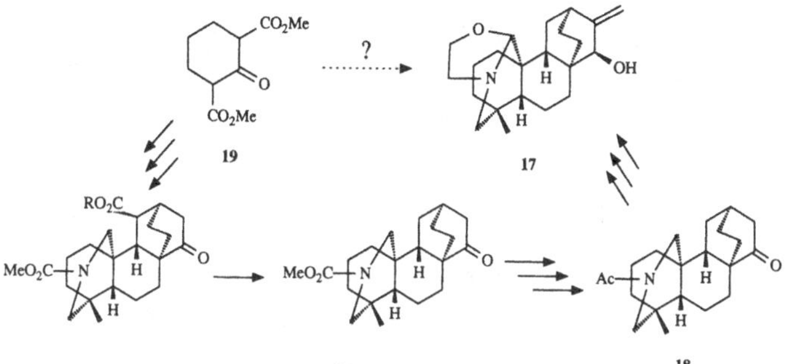

Scheme 6

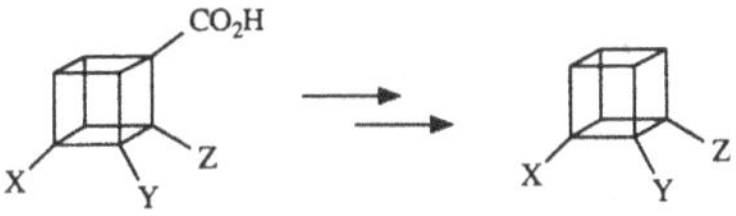

Scheme 7

The special fascination for polycyclic 'cage' molecules and their aesthetic appeal for organic chemists has been quite apparent from the recent literature.[16] During various stages of their work on synthetic routes to cubanes and related systems, Eaton, Hedberg, Della, Castaldi and others have successfully utilized the Barton method, as generalized in Scheme 7.[17-19] Their work has concluded that this decarboxylation process is applicable to a wide variety of bicyclic and polycyclic acids, and that conversions in multi-gram quantities were possible. In view of this fact, and also that this process can quite easily be tolerated by various functional groups and otherwise thermally labile substrates under these mild reaction conditions, it is conceivable to adapt it for large-scale and yet economically viable operation. As part of their studies into the chemistry of bridgehead-substituted polycycloalkanes, Della and coworkers have developed procedures which allow easy access to mono-substituted bicyclo alkanes **20a**–**24b**.[19] They tested the viability

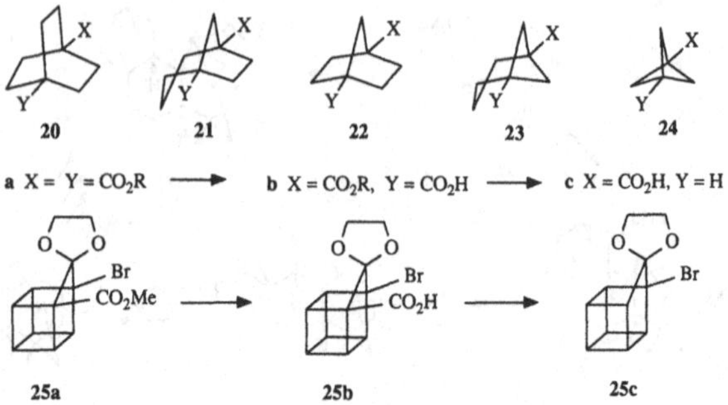

20 21 22 23 24

a X = Y = CO_2R → **b** X = CO_2R, Y = CO_2H → **c** X = CO_2H, Y = H

25a 25b 25c

Scheme 8

of the reductive decarboxylation via the Barton esters of the acids **20b–24b**, Scheme 8. They concluded that the Barton protocol represents a particularly useful method of decarboxylating bridgehead acids where conversion of the half-esters **20b–24b** into acids **20c–24c** was found to proceed in good to excellent yields (57–94 %), despite the fact that generation of the high-energy bridgehead radical is required in each case. As an added example of a strained system, the acetal acid **25b** was subjected to the same decarboxylation sequence and the acetal **25c** was obtained in 73 % yield. Michl has also reported synthetic transformations of bridgehead carboxylic acids **24c** via 1-bicyclo[1.1.1]pentyl bridgehead radicals **24d** generated from the corresponding Barton esters. These radicals (and the related higher staffane systems, $n \geqslant 2$) were employed in studying their reactivity with various substrates, Scheme 9.[20]

Sternbach's work has offered a highly efficient synthesis of the linearly fused triquinane (±)-hirsutene **26**, where intramolecular Diels–Alder reaction of the substituted cyclopenta-

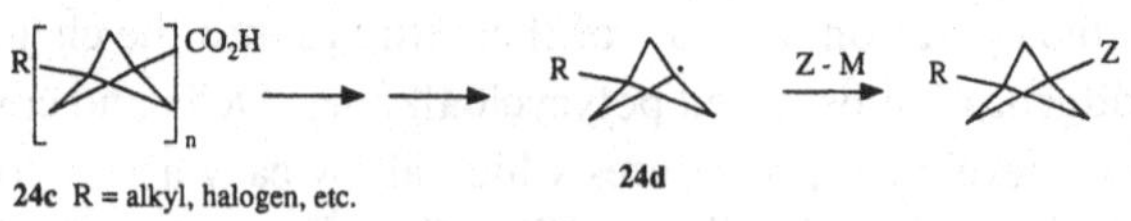

24c R = alkyl, halogen, etc. **24d**

Scheme 9

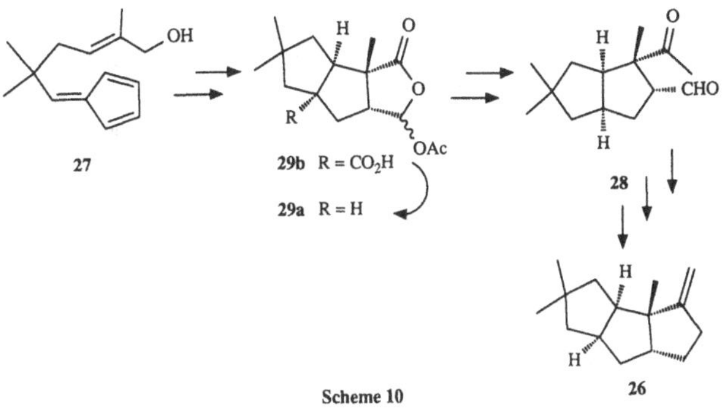

Scheme 10

diene **27**, followed by aldol cyclization of **28** were used for the construction of the requisite ring system. The bridgehead carboxylic group in **29b** was reduced to **29a** using the Barton protocol, Scheme 10.[21]

Magnus's studies on the synthesis of the antitumor *bis*-indole alkaloid vinblastine required development of a straightforward method of making the tetracyclic amine **30** in both the antipodal forms starting with the corresponding tryptophan. The Barton decarboxylation procedure was quite useful in preparation of this indole alkaloid as shown in Scheme 11, and the method proved very compatible with the other functionalities present.[22] During their investigations on the synthesis of (+)-15-(*S*)-prostaglandin A_2 via a facile retro-Diels–Alder approach, Grieco and colleagues have successfully utilized

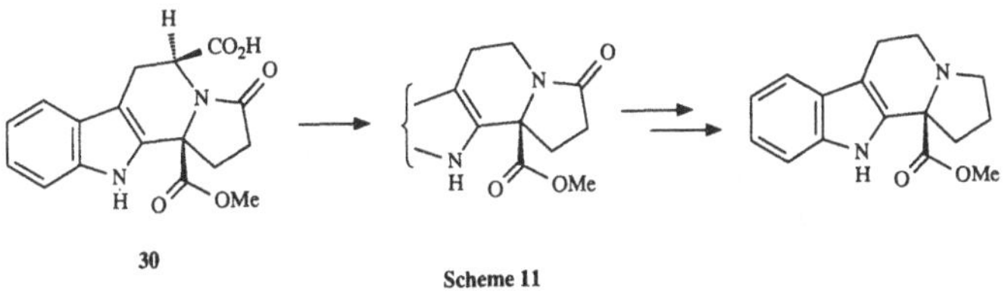

Scheme 11

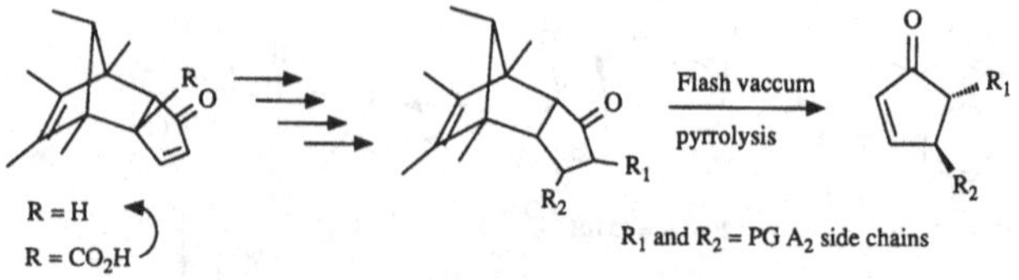

R_1 and R_2 = PG A_2 side chains

Scheme 12

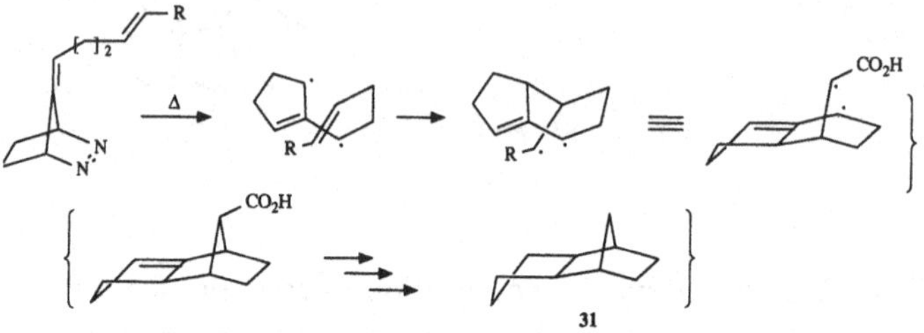

Scheme 13

Barton protocol to remove the bridgehead carboxyl group as shown in Scheme 12.[23]

During their studies on the stereochemistry of intramolecular 1,3-diyl trapping reaction as depicted in Scheme 13, Little and collaborators utilized the reductive decarboxylation in preparing the required compound **31**.[24] The usefulness of Barton decarboxylation was also realized by Braekman and colleagues during their studies on ichthyotoxic sesterterpenoids in providing the needed methyl ketone **32** from the carboxylic acid **33**.[25] Helmchen has developed an easy route to a stable

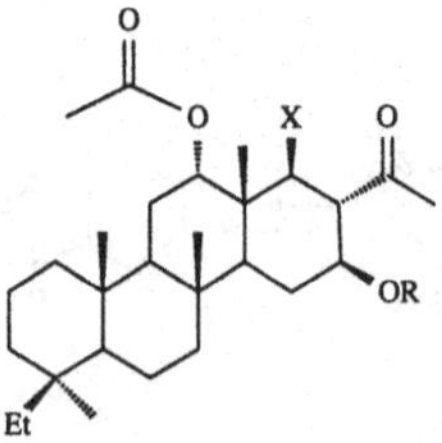

33 X = CHO → CO_2H
32 X = H

precursor **34** for the synthesis of antibiotic sarkomycin **35**, via asymmetric Diels–Alder approach. The unwanted CO_2H in **36** was easily reduced by photolysing the corresponding thiohydroxamate in the presence of ^tBuSH, Scheme 14a.[26] As exemplified in Scheme 14b, Rapoport utilized Barton protocol in reducing the undesired CO_2H in **37** (R = –H) to give **38**, a useful precursor for the synthesis of a potent nerve-depolarizing agent (–)-anatoxin **39**.[27]

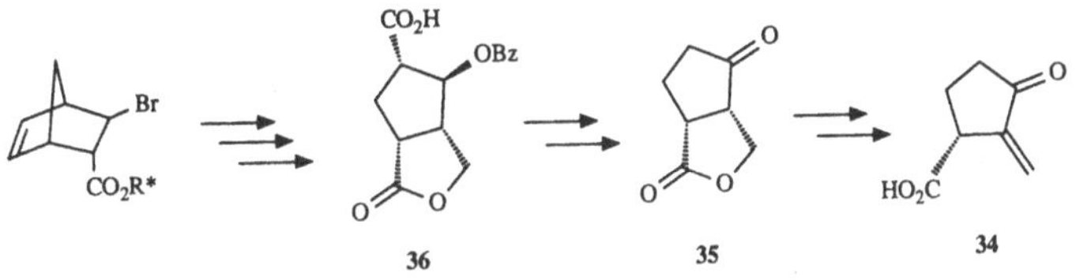

Scheme 14a

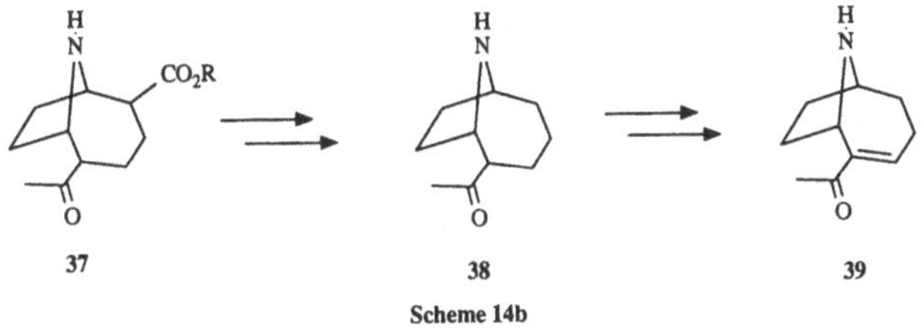

Scheme 14b

In several reports, Crich has shown that the stereochemical outcome of an intermolecular radical reaction is not easily predictable except in the case of glycos-1-yl radicals. Glucosyl radicals, in which the radical center is the anomeric carbon atom, yield predominantly α-substituted products. The selectivity of the glucosyl radical is independent of whether the radical precursor is an α or β anomer, and the stereoelectronic factors associated with the corresponding reactions have been addressed at length by Giese and others.[28] The 1-alkoxyglycos-1-yl radicals **40** in Crich's studies were generated via the

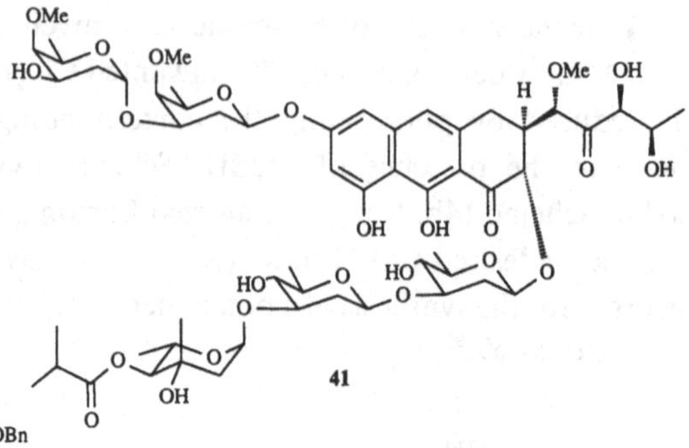

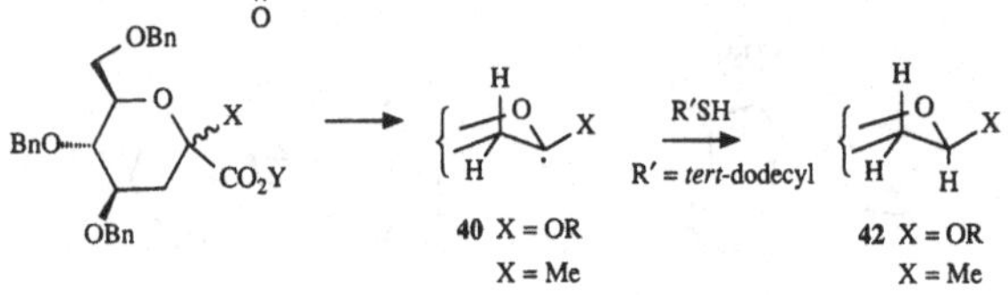

Scheme 15

corresponding Barton esters and the hydrogen atom transfer (from the tert-dodecyl mercaptan[29a]) took place stereoselectively as predicted, providing the corresponding 2-deoxy-β-D-glucosides, Scheme 15. Formation of β-glycoside linkages with 2-deoxy-β-D-sugars is desirable due to their wide occurrence in nature (e.g. olivomycin A[29b] **41**, a member of the aureolic acid group of antitumor antibiotics). 2-Deoxy-β-*C*-pyranosides (X = alkyl, aryl, etc.) were also prepared via this protocol where the hydrogen atom transfer took place stereoselectively during reductive decarboxylation.[29c,d] Methyl ulosonate-*o*-methyl glycoside **42** (X = Me), for example, was synthesized by the formation of the corresponding Barton ester and its subsequent photo-initiated decarboxylation in the presence of *tert*-dodecyl mercaptan giving a 10:1 β:α mixture. Different derivatives (X = *o*-methyl, *o*-*p*-cresyl, *o*-3β-cholestanyl, etc.) were tested and they all gave similar selectivities.[29b–d] In these reports, Crich and coworkers have demonstrated the use of their general procedure for the synthesis of 2-deoxy-β-D-glycosides, and the factors associated with stereoselective atom transfer process at the anomeric radical center during the reduction of the PTOC esters of the parent carboxylic acid have been discussed.[29b–d]

The generality and mildness of the decarboxylative halogenation (Barton alternative to Hunsdiecker reaction) was invoked in light of natural product synthesis by Dauben and coworkers. This radical chain reaction was subjected to milder reaction conditions by photo-initiation at ambient temperatures and this modified light induced reaction gave better yields than the previous thermal process.[30a] Stofer and Lion demonstrated this point further by employing the Barton alternative to the Hunsdiecker reaction in the formation of sterically demanding tertiary chloro derivatives RR′R″CCl from the corresponding carboxylic acids in excellent yields.[30b] This reaction was also found to be useful in the preparation of 1,4-dihalo- bicyclo[2.2.1]heptanes **43c**, from the corresponding carboxylic acids **43a** or **43b**, as shown in Scheme 16a.[31] These

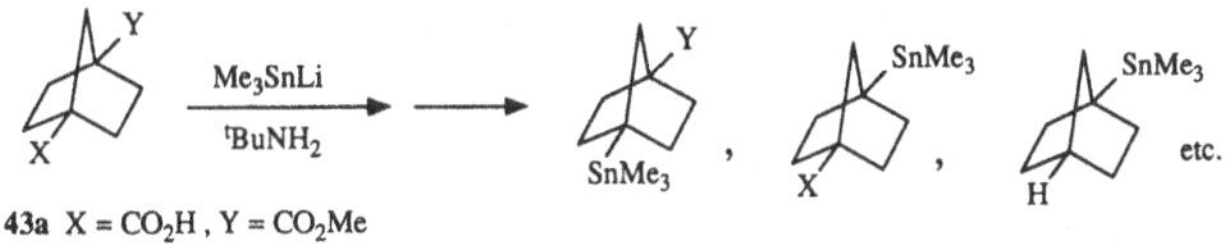

Me_3SnLi, $^tBuNH_2$

43a $X = CO_2H$, $Y = CO_2Me$

43b $X = Cl$, $Y = CO_2Me$

43c $X = Y = Br$ or I

Scheme 16a

dihalo compounds were subjected to trimethylstannylation, and from the resultant product distribution, Adcock has revealed important mechanistic definitions (radical vs polar pathways) for bridgehead nucleophilic substitution reactions.[31]

Vogel and coworkers have described the synthesis of *anti*-1,6:7,12-*bis*methano[14]annulene **44** for studies of its π-electron structure.[32] Incorporation of AIBN into decarboxylative bromination of vinylogous carboxylic acid **45** via Barton esters increased the efficiency of this reaction, Scheme 16b. Harvey

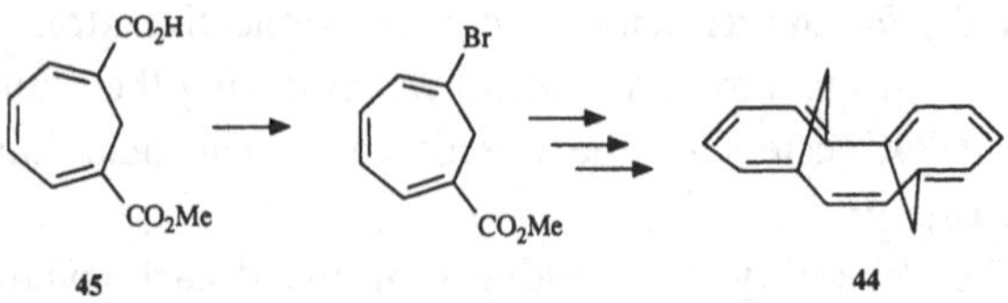

Scheme 16b

has reported efficient general synthetic approaches to polycyclic aromatic molecules based on enamine chemistry, where the requisite 1-(bromomethyl)acenaphthene **46a** was made from 1-acenaphthenyl acetic acid **46b** by the Barton protocol as depicted in Scheme 17a.[33a] During their investigations on the synthesis and reactivity of optically active 1,3-diols (synthesized from dimethyl 3-hydroxyglutarate **47a**), Tamm and coworkers found that the transformation of the half ester **47b** to the corresponding chloro derivative **47c** was only possible employing the radical chemistry of thiohydroxamate, as shown in Scheme 17b.[33b]

Fleet and colleagues have investigated various synthetic approaches for the formation of oxetane nucleosides of which oxetanocin **48** is an important and desirable member.[34] The oxetane carboxylic acids **49a** are helpful precursors for α-chlorooxetanes **50** via **49b**, which in turn provide oxetane

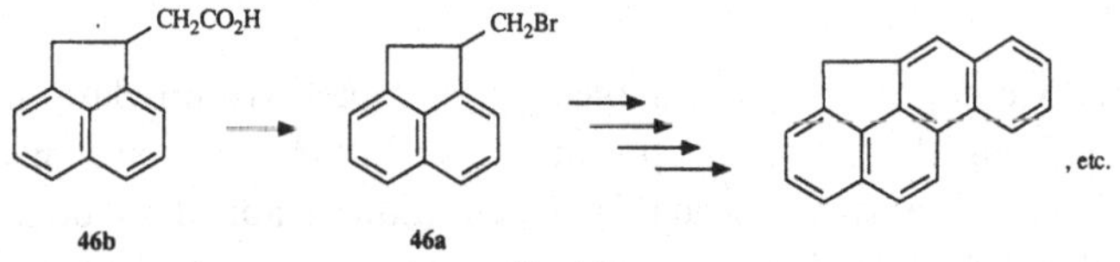

Scheme 17a

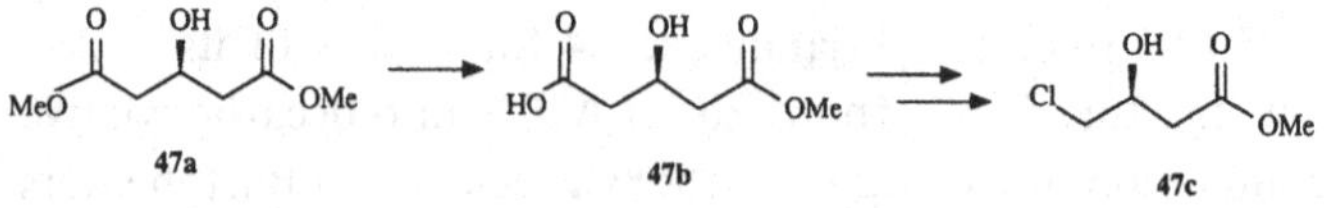

Scheme 17b

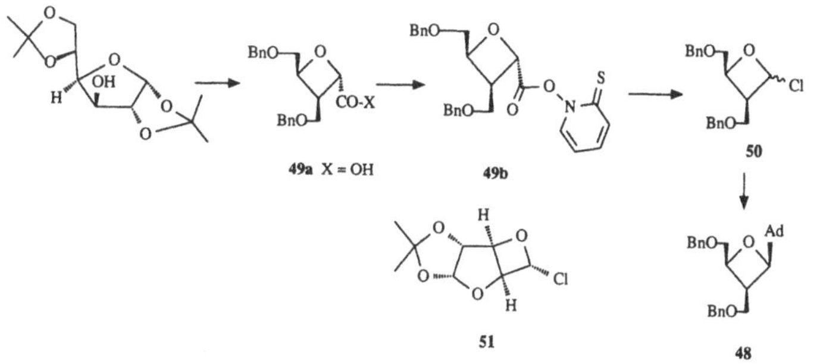

49a X = OH

49b

50

51

48

Scheme 18

nucleosides quite readily.[34] The Barton alternative to the Hunsdiecker reaction was the key step in the preparation of 3,5-anhydro-5*R*-chloro-1,2-*o*-isopropylidinexylofuranose **51** (a stable α-chlorooxetane) and several other α-chlorooxetanes, Scheme 18.[35] Model studies have revealed remarkable information on this class of powerful anti-viral drugs.

α-Alkyl-α-(halomethyl)cycloalkanones have served as useful intermediates in diterpene, alkaloid and sesquiterpene synthesis. Bromocyclobutanones **52** needed for the construction of copaenic sesquiterpenes or as intermediates for a potential ylangene synthesis, are described by Wenkert where the decarboxylative bromination was favorably employed as shown in Scheme 19.[36] In the first enantioselective synthesis of fortamine **53a** the 1,4-diaminocyclitol moiety of Fortimicin A, (deoxyaminoglycoside antibiotic), Hunsdiecker type reactions were found uniformly unsuccessful, as were Baeyer–Villeger type

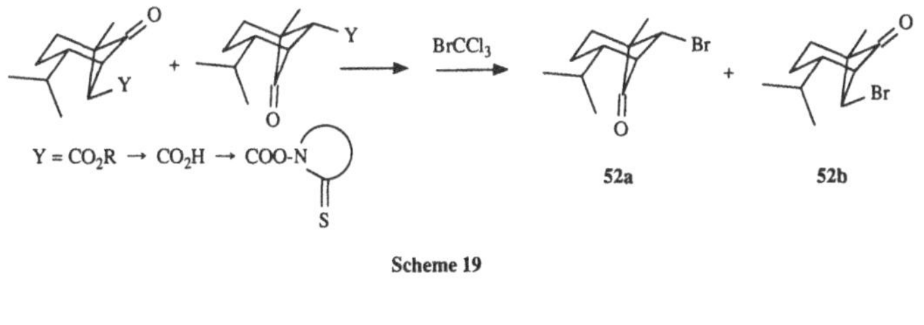

$Y = CO_2R \rightarrow CO_2H \rightarrow$ COO-N

$BrCCl_3$

52a

52b

Scheme 19

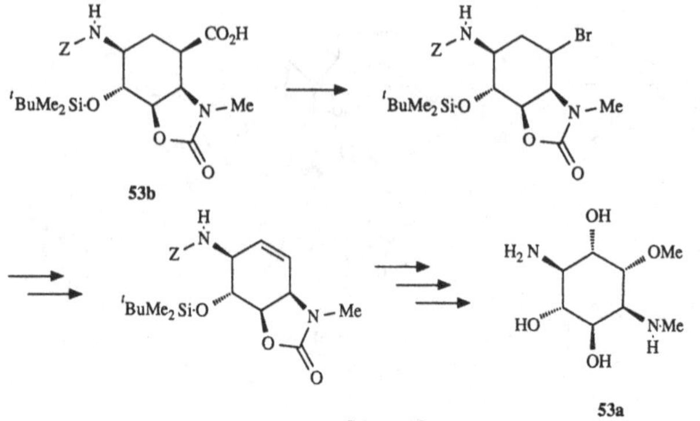

Scheme 20

oxidation pathways to prepare **53a**. Finally, the Barton ester formation, followed by radical bromination, was the only alternative found to be suitable by Ohno and Kobayashi for the transformation **53b** → **53a** as shown in Scheme 20.[37]

Eaton has reported efficient methods for the preparation of substituted iodocubanes from the corresponding carboxylic acids via the Barton alternative in a high yielding radical decarboxylative iodination reaction, where CF_3CH_2I was used as the source of an iodine atom (Scheme 21a).[38] In Szeimies's studies on the reaction of 1,6-dihalohomocubanes **54a**, obtained via Barton decarboxylation in the presence of an appropriate halogen atom transfer agent, and *tert*-butyl lithium after workup gave 1-^{t}butyl homocubane **54b**, where quite

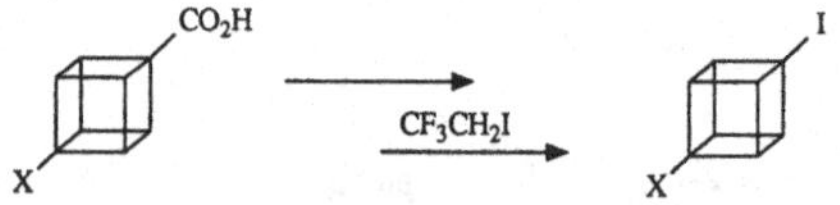

Scheme 21a

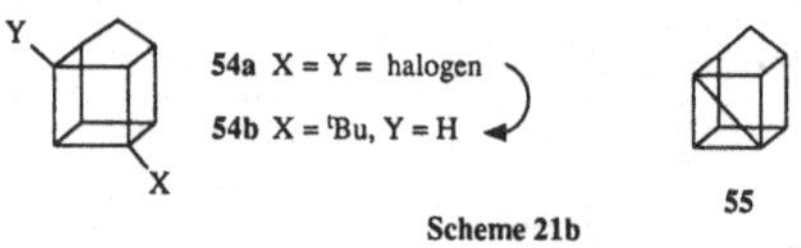

54a X = Y = halogen
54b X = ^{t}Bu, Y = H

Scheme 21b

interestingly, compound **55**, Scheme 21b, was found to be a short-lived reactive intermediate.[39] Cava has recently described a synthesis of **56**, the benzannelated analog of the A unit of antibiotic CC-1065 (one of the most potent antineoplastic agents). The decarboxylative chlorination was applied to obtain **57a** from **57b** in good yields, Scheme 22.[40]

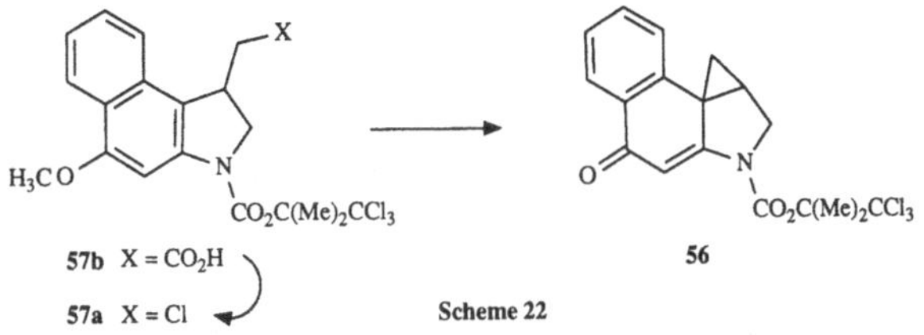

H_3CO X N $CO_2C(Me)_2CCl_3$ O N $CO_2C(Me)_2CCl_3$

57b X = CO_2H

57a X = Cl

56

Scheme 22

Free radical alternatives to carbon–carbon bond forming reactions, via intra- (to a large extent) and intermolecular (to a lesser extent) radical addition are gaining respect. The study of understanding and exploiting substituent effects in radical chemistry is quintessential for radical chemistry as a whole. Crich and Bloodworth have further expanded such substituent and polar effects by showing that the *cis*-cyclooct-4-enyl radical **58a** is reluctant to rearrange. This radical has an option of going to **59** via transannular cyclization or to **60** (unrearranged). The radical was generated both photochemically and thermally from the corresponding *o*-acyl thiohydroxamate and was trapped by CCl_4, 1,1-diethylpropane-1-thiol, **1a**, and oxygen, Scheme 23. The rate of transannular cyclization of this parent radical was found to be relatively slow (and took place only under forcing conditions). However, it could be significantly accelerated by the addition of ring substituents, an effect which is analogous to that found in the cyclization of hex-5-enyl radical to cyclopentyl methyl system.[41] Crich has also utilized this approach to prepare labelled *cis*-cyclooct-4-enyl-hydroperoxide **60d**, as shown in Scheme 23.[41] This type of

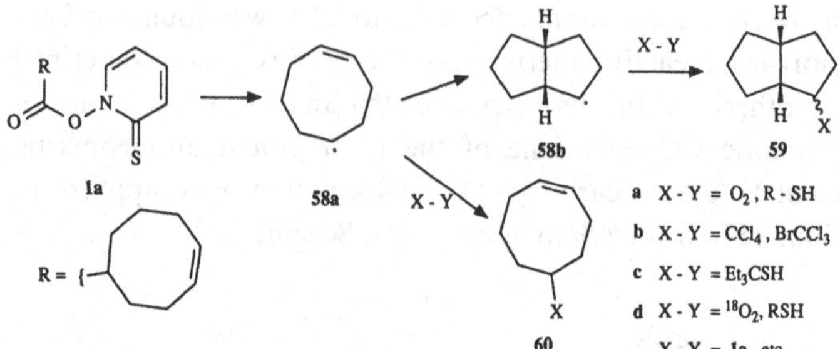

Scheme 23

unsaturated hydroperoxide is quite a useful precursor for the study of reactive intermediates like cyclic peroxonium ions.

Zard and collaborators in their investigations in relation to the synthesis of ring A of nogalamycin-type anthracyclinone antibiotics, found that a sequential radical decarboxylation, addition and cyclization process could be considerably improved (especially the radical addition to olefins) by simultaneous heating and irradiation of the reaction mixture.[42] Compound **61** was obtained in 64 % yield in one pot operation from the acid chloride **62a**, which is readily available from *β*-angelica lactone **63**. This was a considerable improvement over the attempts when the mixture was irradiated at ambient temperature, Scheme 24.[42] In a study involving selective remote functionalization in the longifolene series, Ourisson, Zard and others, demonstrated an interesting 1,5-transannular hydrogen shift, Scheme 25.[43] Radical decarboxylation of **64a** available from isolongifolinic acid **64b** afforded, in good yield, the

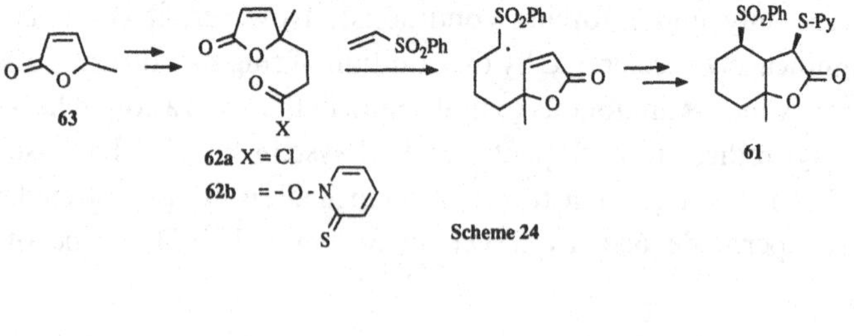

Scheme 24

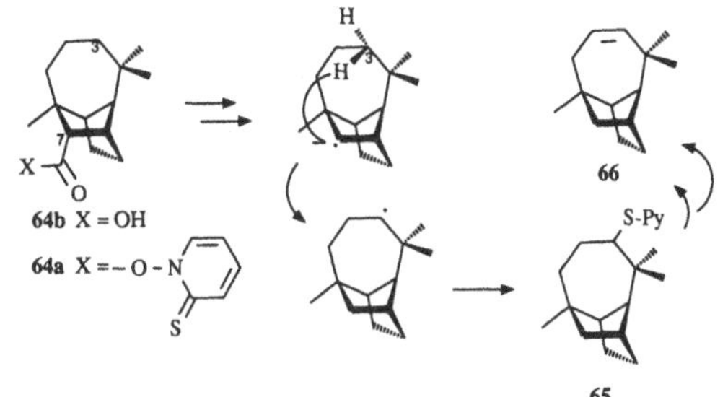

64b X = OH

64a X = –O–N

66

S-Py

65

Scheme 25

thioether **65** through an amazingly clean rearrangement as depicted in Scheme 25.[43] This unique H-atom shift was ascribed as being due to severe steric strain for the radical at the C-7 position, the close proximity of the C-3 hydrogen atom and the non-planarity of this 2-norbornyl radical which is known[44] to be inherently more reactive. The thioether **65** was further converted in almost quantitative yield into olefin **66** (an important member of the longifolene family) through sequential oxidation–elimination.[38] A decarboxylative chalcogenation–elimination sequence was employed as an effective last step by Reich and collaborators, in the preparation of the 5,6-arene oxide of 3,3′,4,4′-tetrachlorobiphenyl **67** as exemplified in Scheme 26.[45]

Cyclopentane ring annulation by a 3+2 radical cycloaddition reaction of 3-alkenyl type radicals **68** and **69** to electron-deficient olefins like **70–71**, represents a useful methodology for the construction of polysubstituted cyclic molecules **72**, **73** etc.[2-4, 46-48] Čeković and coworkers have exercised

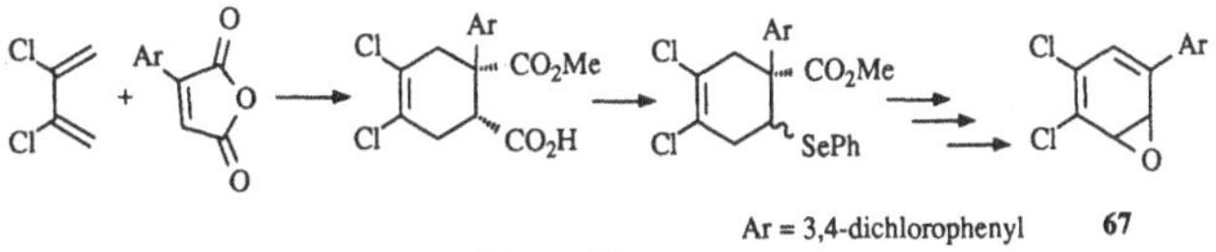

Ar = 3,4-dichlorophenyl **67**

Scheme 26

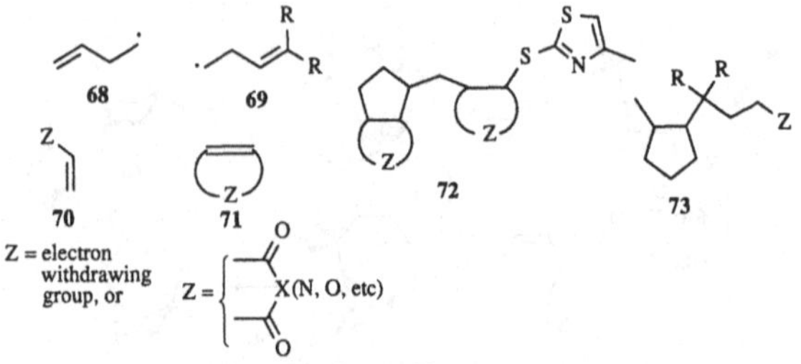

68 69 70 71 72 73

Z = electron withdrawing group, or Z = X(N, O, etc)

Scheme 27

this option quite favorably using Barton esters as demonstrated in Scheme 27.[49] High regioselectivity was achieved in the addition of 3-alkenyl radical **68** to electron-deficient olefinic bond **70**, as well as in 5-*exo*-cyclization of the intermediate 5-hexenyl type radical. By thermally or photochemically induced decomposition of *o*-acyl derivatives of *N*-hydroxy-pyridine-2-thione or other thiohydroxamic acids in the presence of an excess of radicophilic olefins, 2-vinyl cyclopentane derivatives **74** were also obtained, Scheme 28a.[49c] In a similar manner, 3-*exo-trig* type cyclization of the 5-phenylthio-3-pentenyl radical to afford vinylcyclopropanes as shown in Scheme 28b, was also accomplished.[49] This efficient sequence of addition/cyclization/elimination reactions is mediated by a phenylthiyl radical; an effective chain transfer agent.

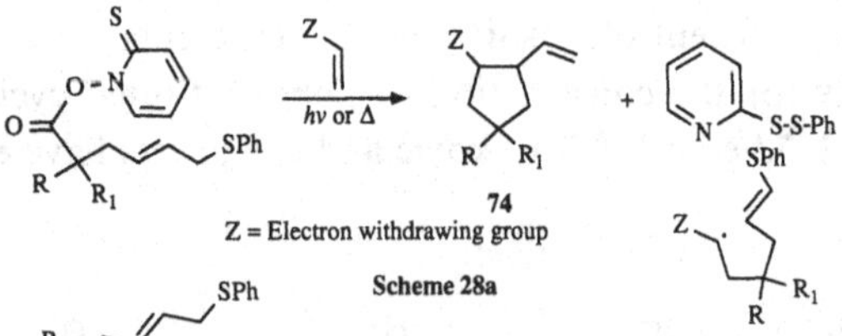

$h\nu$ or Δ

74

Z = Electron withdrawing group

Scheme 28a

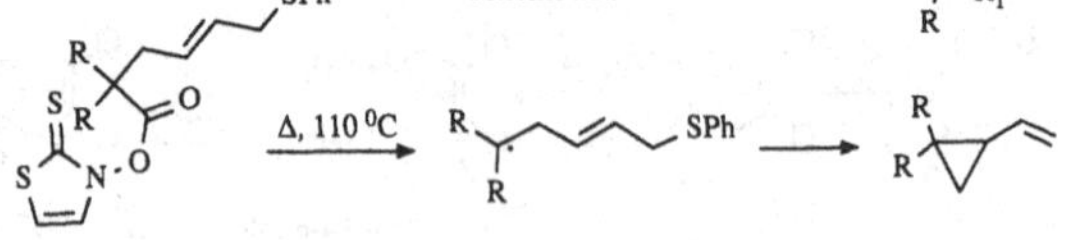

Δ, 110 °C

Scheme 28b

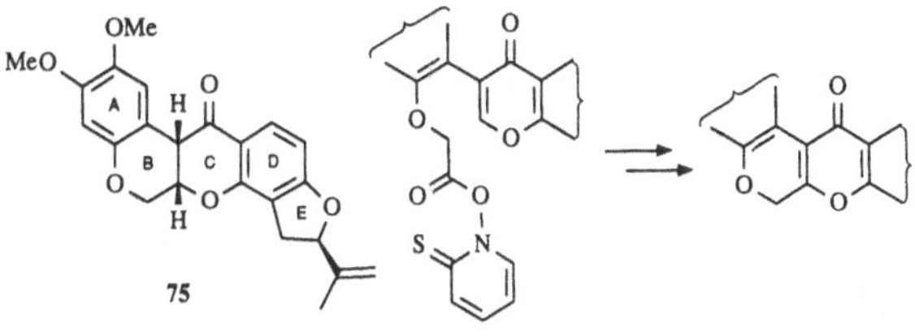

75

76

Scheme 29

Exploitation of decarboxylative radical addition reactions in Whiting's model studies on the biosynthesis of rotenone **75**, have revealed a remarkable model for an enzyme mediated reaction as shown in Scheme 29.[50] The mechanism of biological C–C bond formation from methoxyl groups in the biosynthesis of rotenone **75**, has been postulated via an aryloxyalkyl radical cyclization.[50] The biomimetic cyclization was accomplished by aryl(methylene)oxyl radicals generated via the corresponding Barton ester **76**, which underwent 6-*endo-trig* cyclization to form the rotenoid ring B, Scheme 29.[50] This approach has also been utilized for the construction and biosynthesis of benzodihydropyrans and furan ring systems, as depicted in Scheme 30.[51] Whiting's work has further supported an unusual biosynthetic reaction where a cofactor mediated addition of an

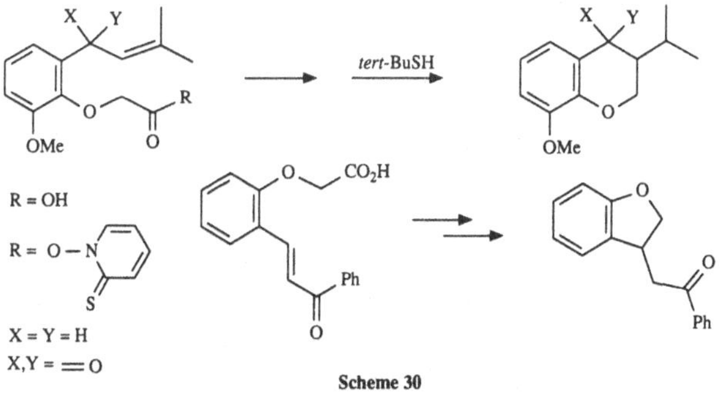

R = OH

R = O–N

X = Y = H

X,Y = =O

tert-BuSH

Scheme 30

aromatic methoxy group to an appropriately substituted double bond is believed to be the key step in the construction of a number of diverse natural products.[52]

Togo and Yokoyama developed a general and efficient method for the synthesis of C-nucleosides employing radical coupling pathways.[53a, b] Several of these C-nucleosides have been synthesized by ionic pathways but they require many steps and suffer a lack of generality. The thiohydroxamates derived from pentose or 2-tetrahydrofuryl carboxylic acid, Scheme 31, gave the corresponding C-nucleoside derivatives **77** in the presence of an appropriate heteroaromatic compound.[53b]

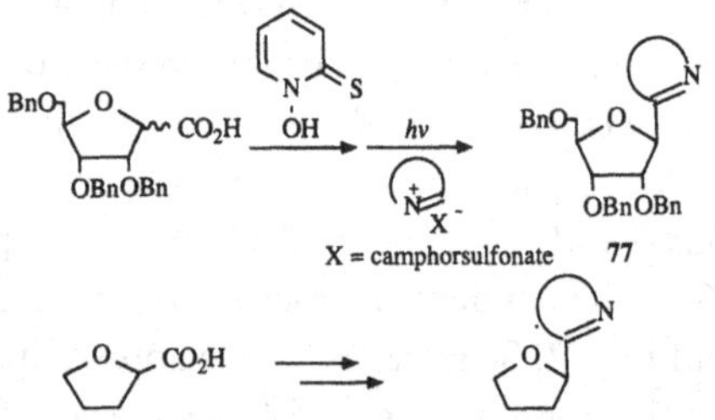

X = camphorsulfonate

Scheme 31

Asymmetric induction and stereoselectivity are the two terms that have not been properly addressed in terms of intermolecular radical reactions.[54] Recently, however, high levels of asymmetric induction have been observed for trapping with alkenes containing stereogenic centers. For example, low temperature carbon radical addition of α,β-unsaturated amides **78b–c** containing the C_2-symmetrical *trans*-2,5-dimethylpyrrolidine, was achieved by Porter and coworkers in high yield giving **79b–c**, with unprecedented diastereoselectivity (> 99 % de in **79c**) in the decarboxylative radical addition reaction upon photolysis of Barton esters **80a**, Schemes 32a, b.[55] Use of the olefin **78c** substituted with three electron withdrawing

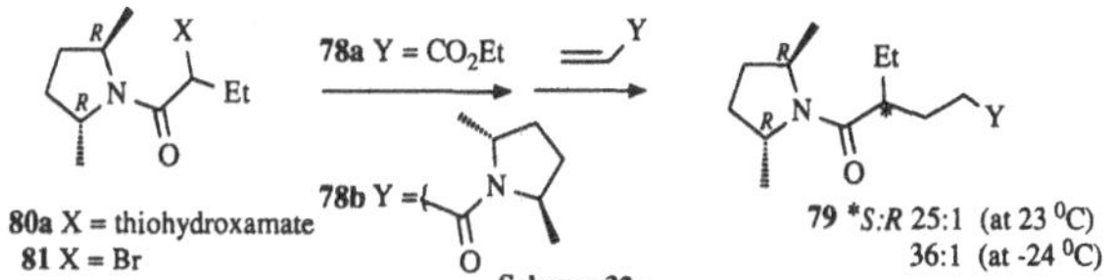

78a Y = CO_2Et

78b Y =

80a X = thiohydroxamate
81 X = Br

79 *S:R 25:1 (at 23 °C)
36:1 (at -24 °C)

Scheme 32a

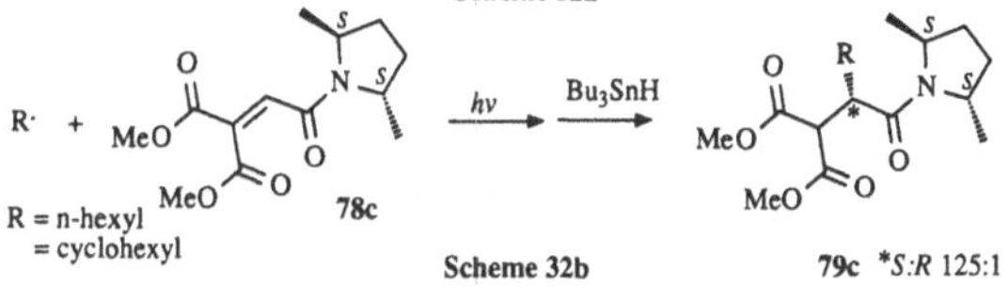

R = n-hexyl
= cyclohexyl

78c

hv Bu_3SnH

79c *S:R 125:1

Scheme 32b

^{t}Bu· -35 °C, **78b**

82a 14 : 1 **82b**

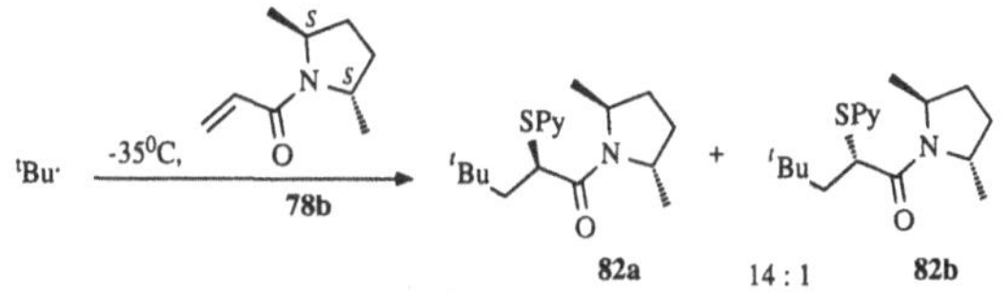

Scheme 32c

groups offers synthetic versatility and regiospecificity, Scheme 32b.[55b] Porter and coworkers have also made comparison studies in this case employing various approaches (based on **80** and **81**) to generate radical species. At sub-ambient temperatures, Barton protocol offered significant improvement in stereocontrolled intermolecular radical addition, where the tin method or the mercury method offered only moderate selectivities for **79**.[55] On the similar substrate **78b**, Giese showed that atom/group abstraction by the chiral adduct radical, also occurs stereoselectively.[56] The adduct radical was generated after the first addition of *tert*-butyl radical, which was procured from the corresponding Barton ester, and the products **82** were obtained in a 14:1 ratio as shown in Scheme 32c.[56] The C_2 symmetry of the pyrrolidine moiety is apparently responsible for this extraordinary selectivity observed in these amide cases. A conformational restriction around the C–N bond renders the two methyl groups of the amine component shielding the two sides of the alkene to different extents.[55b] A detailed discussion on this subject has recently been reported.[55c] Asymmetric induction in the radical addition reaction to C–C multiple bonds containing a distal chiral center has also been addressed recently. Crich showed that moderate diastereoselectivity could be obtained in the addition of alkyl radicals to chiral acrylate esters **83** as exemplified in Scheme 33.[57] Diastereofacial

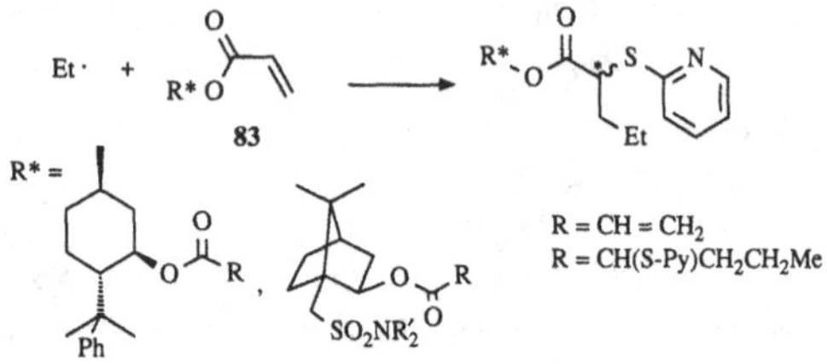

Scheme 33

differentiation was moderate due to the fact that the controlling center was further away. These results make it clear that radical reactions in acyclic systems can indeed be stereoselective. The interesting 1,2 and 1,4 and to a lesser extent acyclic stereo-induction in this type of radical addition reaction has opened new avenues for future work in this area. Beckwith showed that the cyclic exo-methylene compounds **84** undergo diastereo-selective free radical addition when treated with alkylmercury hydride or alkyl iodide/tributylstannane to give products **85**, Scheme 34.[58] Barton protocol was favorably exploited in this study which offers useful precursors for the enantioselective synthesis of α-amino α-hydroxy acids.[58]

A cyclopropyl radical is a rapidly inverting bent σ radical, incapable of maintaining its configuration. However, elec-

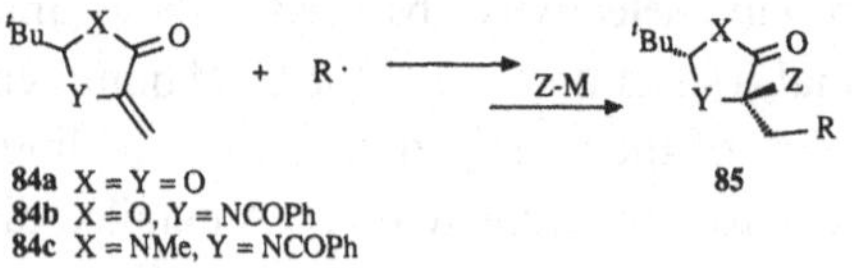

Scheme 34

tronegative substituents such as fluorine or alkoxy in the α-position have been shown by Walborsky and collaborators, to slow down the inversion frequency to the extent that if a very good radical trap is available then the radical can partially or completely retain its configuration.[59] The chiral 1-fluoro-2,2-diphenylcyclopropyl radical **86** generated via the Barton decarboxylation reaction from the chiral acid **87** was used as a probe to evaluate the stereo-controlled atom/group abstraction reaction with a variety of halogen and hydrogen atom donating reagents, Scheme 35.[60] The optical purity of various

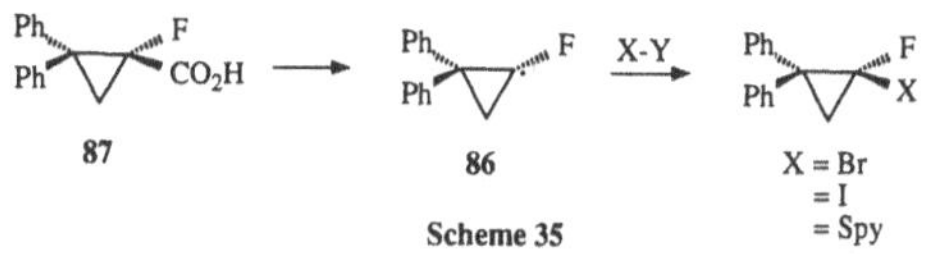

Ph Ph F CO_2H → Ph Ph F X-Y → Ph Ph F X

87 **86** X = Br = I = Spy

Scheme 35

trapped products was measured and the stereo-electronic factors associated with this reaction have been postulated.[60] Brandi and Pietrusiewicz have accomplished radical addition to vinyl phosphine oxide **88a** with an interesting 1,2-stereo-induction of the phosphorus stereogenic center, as depicted in Scheme 36.[61] The use of chiral vinyl phosphine oxide **88a** possessing a sterically demanding mesityl group allowed stereoselectivity as high as 9:1 in **88b**. The conformational preference is exemplified in **88c**, Scheme 36.

Barton esters **1** have been beautifully exploited for the

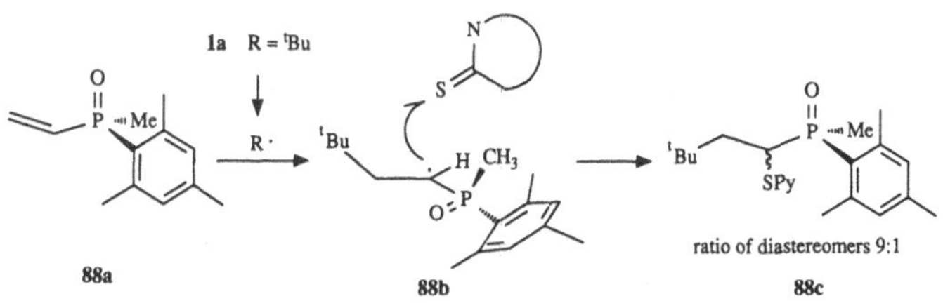

1a R = ᵗBu

O P Me R· ᵗBu N S H CH_3 O P ᵗBu O P Me SPy

ratio of diastereomers 9:1

88a **88b** **88c**

Scheme 36

generation of various heteroatom centered radicals. Newcomb, Zard and Beckwith have shown that alkoxycarbonyloxy radicals such as **89a**, **90a** generated from PTOC carbonates of allyl and homoallyl alcohols cyclize in an *exo-trig*-fashion to give 3-substituted 1,2-diol and 4-substituted 1,3-diol carbonates **89b** and **90b** respectively, as shown in Scheme 37.[62]

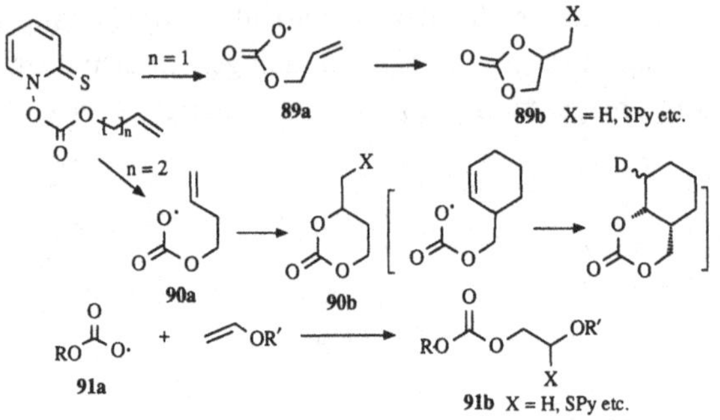

Scheme 37

Beckwith has observed moderate diastereoselectivity in this reaction involving cyclic systems. The cyclizations of allyl and homoallyloxycarbonyloxy radicals are potentially useful as radical based alternatives for an overall oxidation or hydrolysis of a double bond, and also various further transformations of the cyclic carbonates can lead to synthetically useful products.[62b] In contrast, simple alkoxycarbonyloxy radicals **91a** add intermolecularly to ethyl vinyl ether to give, ultimately, carbonates of glycoaldehyde derivatives **91b**, Scheme 37.[62a]

Ovchinnikov's proposal on the basis of valence bond theory had predicted that polyacetylene carrying phenoxy radicals as pendants **92a** (X = H) might become an organic ferromagnet.[63a] Several attempts to synthesize **92a** had failed previously, largely due to recombination and/or disproportionation reactions.[64] In relation to their studies of organic ferromagnets, Iwamura and coworkers have synthesized and studied the photolysis of *N*-(phenoxy-carbonyloxy)-2-thio-

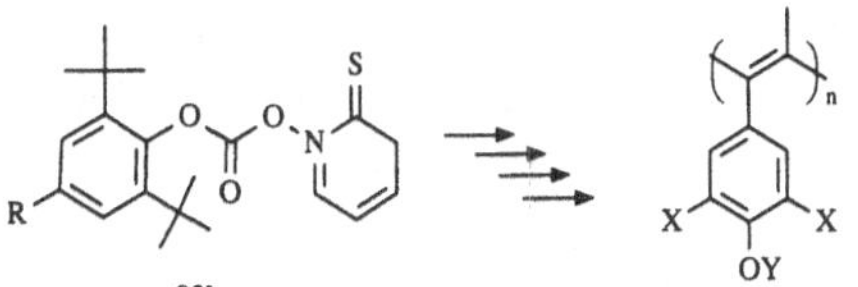

92b

92a X = ^{t}Bu, H; Y = ·

Scheme 38a

pyridone derivatives **92b**, as shown in Scheme 38a. They came to a realization that unimolecular generation of phenoxy radicals in the solid state, where diffusion of free radicals would be restricted, is highly desirable. Their finding showed that the acyl derivatives **92b** are very stable and are readily converted to phenoxy radicals in excellent overall yield without any impurity, Scheme 38a.[63b] The modification of polymers by surface grafting is widely used for altering the surface properties of a preformed polymer film. A principal problem in such chemistry is the synthesis of an initiator site on a solid polymer. As part of their studies on developing new methods to elaborate synthetically functionalized surfaces, Bergbreiter and colleagues thought of employing Barton esters as graft precursors to generate grafting sites with an idea of controlling the degree of polymerization of the graft if they would prove to be useful as initiators. They have kindly informed us of their findings

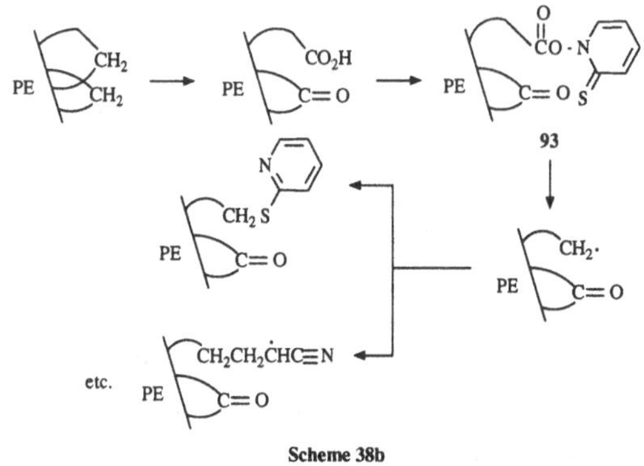

93

etc.

Scheme 38b

which demonstrate the use of this new type of macroinitiator for surface grafting that should be useful with a variety of polymer surfaces containing carboxylic acid groups.[64] As demonstrated in Scheme 38b, the polyethylene surface was appropriately modified to prepare the corresponding thiohydroxamic acid derivative **93**. Subsequent radical chemistry was studied with various radical acceptors and contact angle measurements confirmed that grafting had occurred. ATR-IR and XPS spectroscopic techniques were used to determine the nature and extent of grafting. These studies point to yet another field of application of Barton esters.

Beckwith has shown that alkoxy radicals can be generated by simply heating the *o*-alkyl thiohydroxamate **94** in the presence of Bu_3SnH.[65] The formation of 2-methyltetrahydrofuran **95a** (80 %) from **94a**, as shown in Scheme 39, was used to estimate

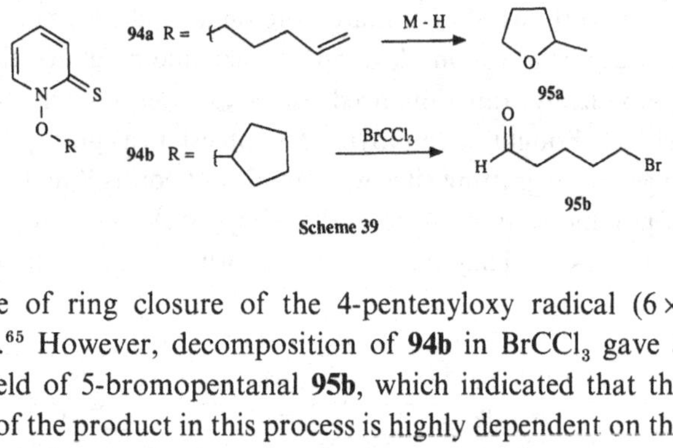

94a R = ; M-H ; 95a
94b R = ; $BrCCl_3$; 95b

Scheme 39

the rate of ring closure of the 4-pentenyloxy radical ($6 \times 10^8\ s^{-1}$).[65] However, decomposition of **94b** in $BrCCl_3$ gave a high yield of 5-bromopentanal **95b**, which indicated that the nature of the product in this process is highly dependent on the type of chain-transfer agent involved.

Newcomb and coworkers have very elegantly demonstrated that *N*-hydroxypyridine-2-thione carbamates **1c** (PTOC carbamates, $X = NR_2$) are directly analogous to the *N*-hydroxypyridine-2-thione esters **1a**. They can be prepared readily from primary and secondary amines. Reaction of these carbamates in a radical chain process gives carbamoyl radicals that decarboxylate to give aminyl radicals, Scheme 40.[66-71] In the

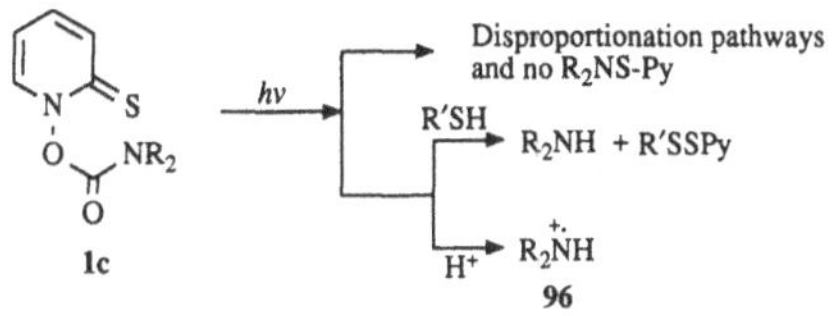

Scheme 40

absence of a hydrogen atom donor, disproportionation of the aminyl radical occurs, leading to the conclusion that aminyl radicals are not thiophilic with respect to the thiocarbonyl group. However, in the presence of a protic acid and a thiol, **1c** reacts in a chain fashion to give aminyl radicals which can be protonated to give highly reactive aminium cation radicals, **96**.

These radical cations, if possessing δ,ε-unsaturation, cyclize to give 5-membered rings **97**, Scheme 41, and thus provide a

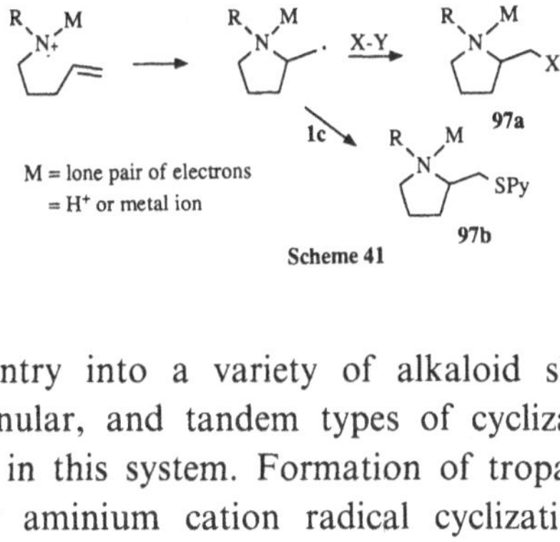

M = lone pair of electrons
= H^+ or metal ion

Scheme 41

rapid entry into a variety of alkaloid skeletons. Simple, transannular, and tandem types of cyclizations have been studied in this system. Formation of tropanes **98** by transannular aminium cation radical cyclization as shown in Scheme 42 is demonstrative of the same. *N*-Alkylcyclohept-4-enaminium cation radical **99** cyclized to give ultimately tropane and substituted tropanes **98** in high yields, Scheme 42. The sequences generated under mild and controlled conditions have also been found to be useful for kinetic studies. Newcomb has shown that aminium cation radicals add intermolecularly as well to the electron-rich double bonds and thus serve as a source of electrophilic nitrogen, Scheme 43. When *N*-allyl

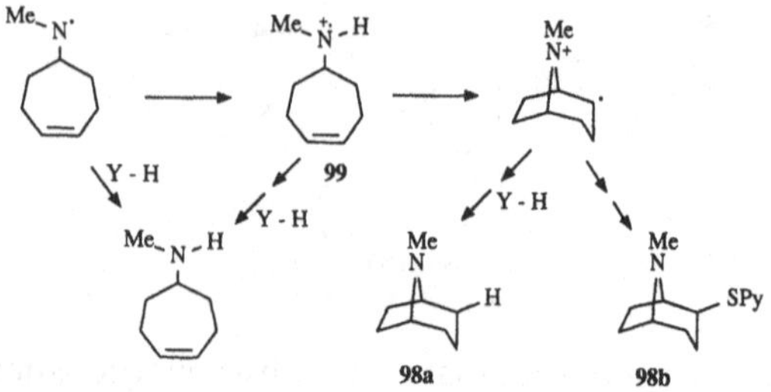

Scheme 42

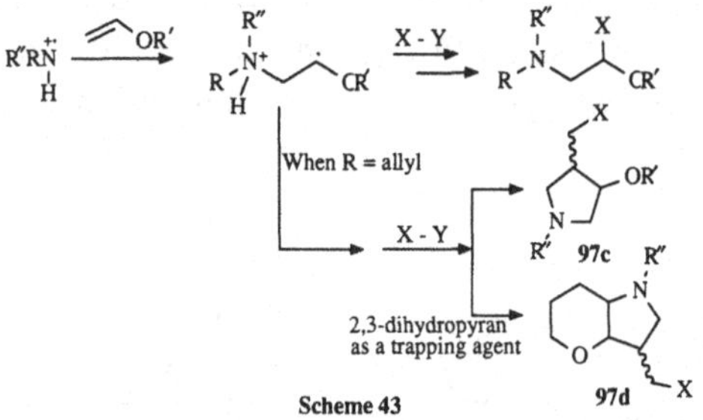

Scheme 43

aminium cation radicals are employed in a sequence, a tandem addition–cyclization occurs to give pyrrolidine products **97c–d** via a 3 + 2 cycloaddition, Scheme 43. A variety of reagents have been evaluated in these reactions by Newcomb for trapping carbon radicals formed from aminium cation radical cyclizations. Due to rapid 'self-trapping' of these radicals by their PTOC carbamate precursors, it does require highly reactive trapping agents. Synthetically useful trapping agents were found to be ^{t}BuSH (radical reduction), self trapping, olefin trapping and also CBr_4, Ph_2Se_2, CCl_4, etc. A typical 5-*exo* radical cyclization followed by trapping with ^{t}BuSH or PTOC carbamates gives rise to a variety of alkaloid skeletons. The ring systems obtained (typically in good to excellent yields) include pyrrolidines **97**, tropanes **98**, perhydroindoles **100**, pyrrolizidines **101**, 9-aza-bicyclo[4.2.1]nonanes **102**, 6-aza-

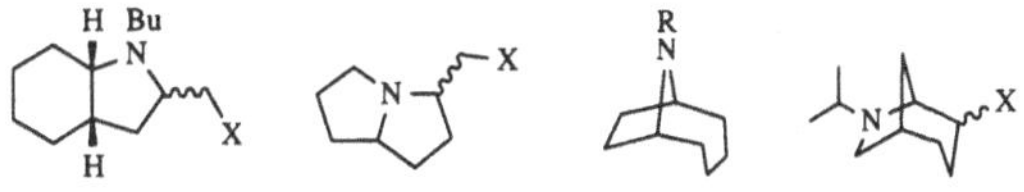

100 101 102 103

bicyclo[3.2.1]octanes **103**. It was found that 6-*exo-trig* and 7-*endo-trig* cyclizations competed in the 6,7-unsaturated systems.[69,70]

Surprisingly, however, monoalkylaminium cation radicals **104** have not been studied well. Newcomb found that they could not be prepared from respective PTOC carbamates due to instability of the precursors, nevertheless, the problem was solved via the isocyanate condensation with a different thiohydroxamic acid **105a**, as shown in Scheme 44. These deriva-

R-N=C=O + 105a → 105b → R - $\dot{N}H_2^{+}$ (104) → etc.

Scheme 44

tives termed as TTOC carbamates **105b**, are more stable and recently their relevant chemistry has been investigated.[72] Another promising adaptation of this approach was conceived and materialized with the generation of the desirable amidyl radicals. Amides were converted via imidoyl chlorides **106** into *N*-hydroxypyridine-2-thione imidate esters, **107**, Scheme 45. These upon photolysis generated the amidyl radicals, **108**. In the presence of δ,ε-unsaturation, cyclized products **109** were obtained as shown in Scheme 45.[73] This chemistry based on nitrogen-centered radicals is attractive not only for synthetic conversions but also appealing due to a display of polarity reversed reaction character.

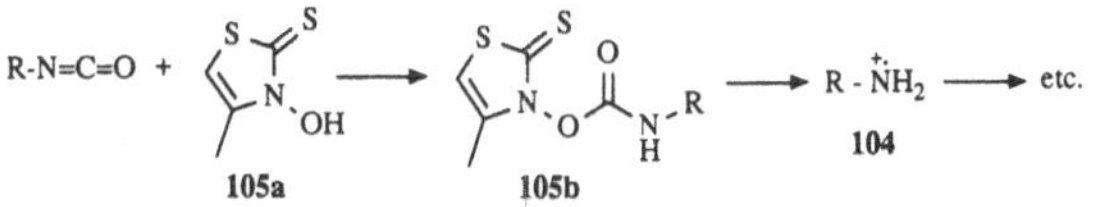

Togo and Yokoyama demonstrated that the decarboxylation reaction of *N*-alkoxyoxalyloxy-2-thiopyridone **110** which was

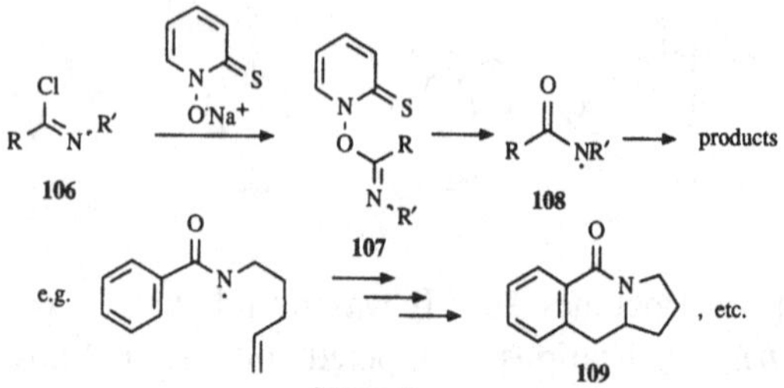

Cl
R N R′
106
N S
O⁻Na⁺
N S
O R
N R′
107
O
R NR′
108
products
e.g.
O
N
O
N
109
, etc.

Scheme 45

prepared by the reaction of alcohol, oxalyl chloride and *N*-hydroxy-2-thiopyridone undergoes facile fragmentation and the resultant radical offers a variety of transformations, as shown in Scheme 46.[74] This reaction was studied both in the presence and absence of olefinic compounds, Schemes 46a, and 46c–d.[74] The same reactions with olefinic and acetylenic

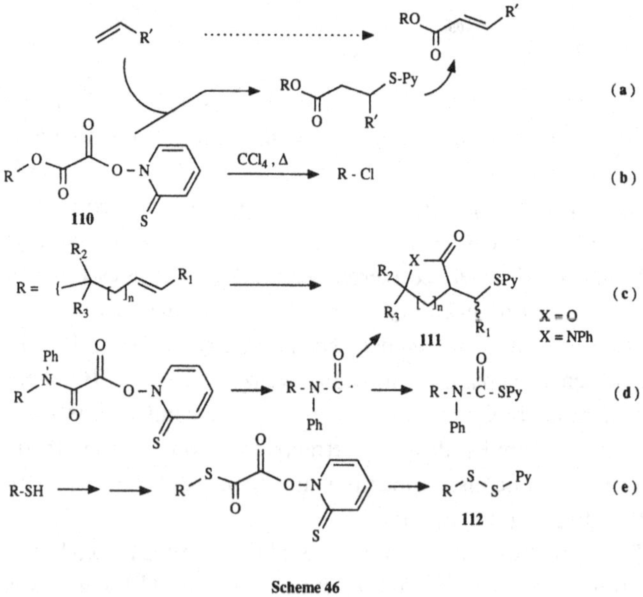

R′
RO
O
R′
RO
O
S-Py
R′
(a)
R O
O
O
O – N
S
110
CCl_4, Δ
R - Cl
(b)
R_2
R =
R_3
n
R_1
X
O
R_2
R_3
n
SPy
R_1
111
X = O
X = NPh
(c)
Ph
R N
O
O
O – N
S
O
R - N - C ·
Ph
O
R - N - C - SPy
Ph
(d)
R-SH
R S
O
O
O – N
S
R S S Py
112
(e)

Scheme 46

alcohols gave the corresponding cyclized products, the γ-lactones and γ-lactams **111**, starting with 3-olefinic alcohols or amines. On the other hand, the unsymmetrical alkyl-2-pyridyl disulfides **112** were obtained by the same reaction with aliphatic thiols, Scheme 46e. Reactions with various primary, secondary or tertiary alcohols and dialkyl amines gave the corresponding thiocarbonates, thiocarbamates respectively. In addition, the applications of this adaptation of Barton decarboxylation provides several alternative manipulations of other functional groups. Crich for example, showed that the conversion of tertiary alcohol into alkyl chloride via the mixed oxalate **110**, Scheme 46b, is quite facile and devoid of any detrimental side reactions usually encountered in the corresponding ionic reactions.[75]

Relatively few methods are convenient for generating hydroxyl radicals besides the well-documented Fenton reaction.[76] Recently Zard and Boivin examined the possibility of radical fragmentation of *N*-hydroxy-2-thiopyridone **113** giving rise to hydroxyl radicals upon irradiation with visible light. They indeed generated the hydroxyl radicals and incorporated them in useful radical chain processes, Scheme 47, involving hydrogen atom abstraction from appropriate sources, R–H, and producing R$^{\bullet}$.[77] Besides the various possible synthetic applications, this method can also serve as a tool for studying many

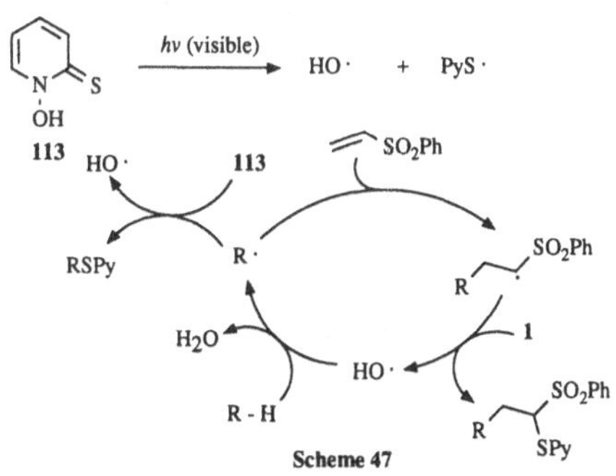

Scheme 47

of the crucially important biological processes.[77] Zard has also reported several synthetic applications of iminyl radicals generated via different approaches. Barton decarboxylation was adapted for this purpose for efficient generation of iminyl radicals in a rather unique fashion as shown in Scheme 48. Homolytic cleavage of N–O bond in the thiohydroxamic acid derivatives upon thermolysis or photolysis has been demonstrated to undergo decarboxylation, double decarboxylation or even rearrangement leading to carbon or heteroatom centered radical species. Zard and colleagues have adapted and modified this idea one step further for the purpose of generating iminyl radicals **114** from the PTOC derivative **1a** (R=-CH_2-O-N=CR′R″), as illustrated in Scheme 48.[77] It is

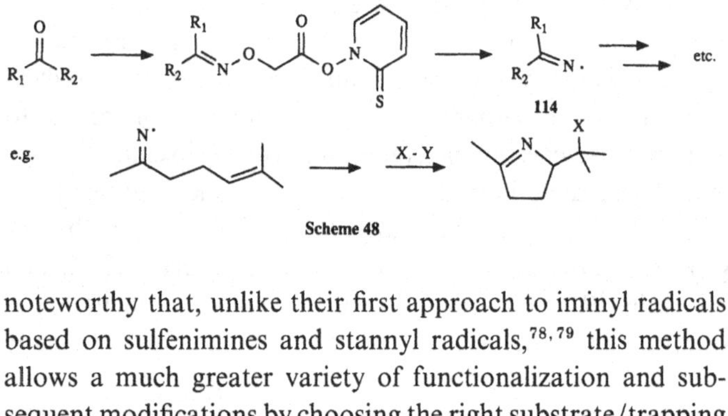

Scheme 48

noteworthy that, unlike their first approach to iminyl radicals based on sulfenimines and stannyl radicals,[78,79] this method allows a much greater variety of functionalization and subsequent modifications by choosing the right substrate/trapping agent. The synthetic transformations **115** to **116** as outlined in Scheme 49 involves an interesting regiospecific opening of cyclobutane ring systems.

Radical-based dephosphorylation via a phosphonyl radical constitutes one formulation for the mechanism of C–P bond cleavage during microbial degradation of organophosphonates.[80] Frost has demonstrated that anhydrides **117** derived from phosphonic acid and thiohydroxamic acid react in a chain sequence with thiols, Bu_3SnH and CCl_4 to give dephosphoryl-

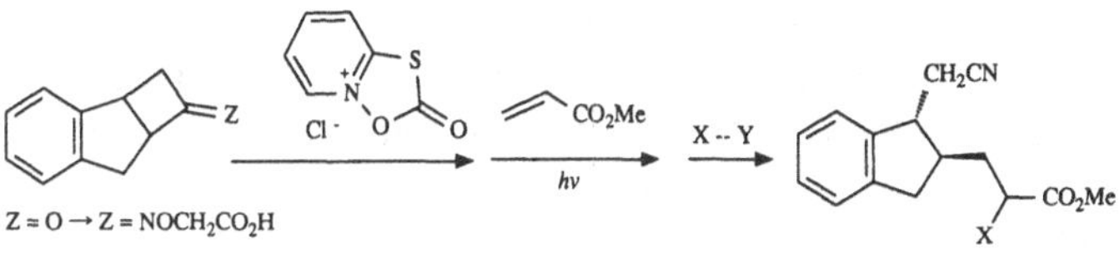

Scheme 49

ated products. For simple primary and secondary alkyl phosphonates, the yields were low, which suggested inefficient fragmentation or competing polar reactions. Allyl or benzyl phosphonates, on the other hand, underwent dephosphonylation in moderate to good yields giving the corresponding reduced species or chlorinated species as shown in Scheme 50.[81] This work opened a new route into the monomeric metaphosphate reaction manifold.

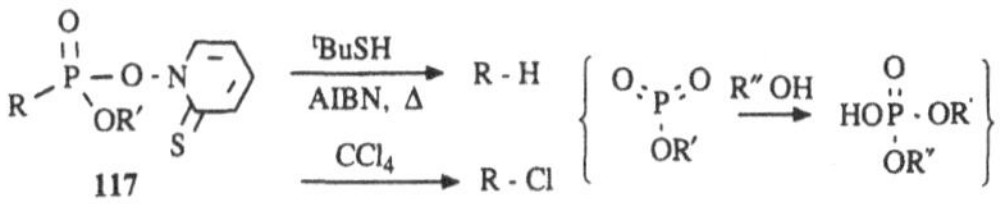

Scheme 50

All these reactions discussed thus far have called upon thermal or photoinitiation techniques for the generation of radicals from Barton esters. An altogether new approach was developed by Dauben and coworkers where the $\cdot CCl_3$ radical generated by ultrasound propagates the chain sequence outlined in Scheme 51.[82]

Investigation of the reaction mechanisms and the deter-

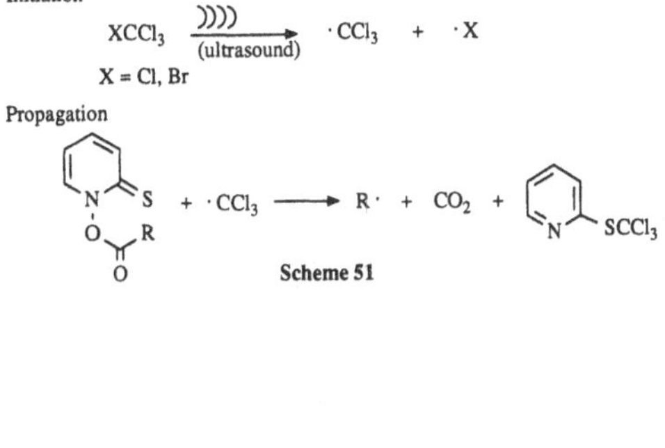

Scheme 51

mination of the relevant kinetic parameters of the radical reactions are very important as they help develop a foundation for future work and offer grounds for predicting the outcome of new reactions. It is very desirable for a successful synthetic methodology based on radical reactions that the overall reaction sequences involve *chain* processes in order to maintain a low concentration of radicals. Chain reactions can be quite complex and because of the large number of steps in a chain reaction, the derivation of kinetic rate laws could be quite intricate. The rate constants for radical reactions can normally be measured directly by use of spectrophotometric or spectroscopic techniques or indirectly by use of competition reactions which are often based on a 'radical clock' reaction. The radical reaction used for comparison must have a well-measured rate constant, and a relatively clean source of radicals must be available. The Barton esters have been shown to be one such clean and convenient source of C, N, S or O centered radicals. Newcomb, Ingold, Beckwith and others have utilized several of these techniques discussed thus far in measurements of rates of competing reactions, where the option of product oriented analyses offers a direct probe into the kinetic parameters.[83–90] It would not be possible here to discuss all their important findings of kinetic parameters. However, it is quite appropriate to state that, in the use of benzene selenol as a trapping agent, Newcomb has been able to accomplish an ultimate kinetic resolution of $\approx 1 \times 10^{-12}$ s, at room temperature in a H-atom transfer trapping of cyclopropyl carbinyl radical.[89,90] This is the first report of its kind where the rate of a diffusion-controlled reaction has been probed using an indirect technique.

The synthetic utility of any reaction is normally viewed in the light of ease of preparation of starting materials, overall yields and number of steps involved in the process and its large-scale adaptability. It is interesting to note that in their synthetic studies on the potent immunosuppressant FK 506 (Tsukubaenolide) which is currently the subject of intense pharmaceutical interest, Kocieński and Donald required a large scale synthesis of an important intermediate **118**. This compound needed for the preparation of **119**, the C24–C34 segment of FK 506 was synthesized via the Diels–Alder reaction on a 20 g scale, as shown in pathway A, Scheme 52.[91] However, owing to the increased number of steps, racemization during the hydrolysis, and mediocre yield in the final oxidation step, an alternate route needed to be developed. Readily available *meso*-diester **120** was hydrolyzed using pig liver esterase (PLE) to monoester **121** followed by Barton protocol gave the requisite ester **122** on 61 g scale in overall 95 % yield over three steps as depicted in pathway B, Scheme 52. This application is exemplary of the synthetic utility of radical chemistry based on thiohydroxamic acid derivatives. Moreover, it is interesting to note that all these different methodologies and applications and kinetic adaptations, based on the radical chain reactions of Barton esters and related systems, have been developed in a relatively short amount of time, which, in turn, speaks of the potential and versatility of these radical methods.

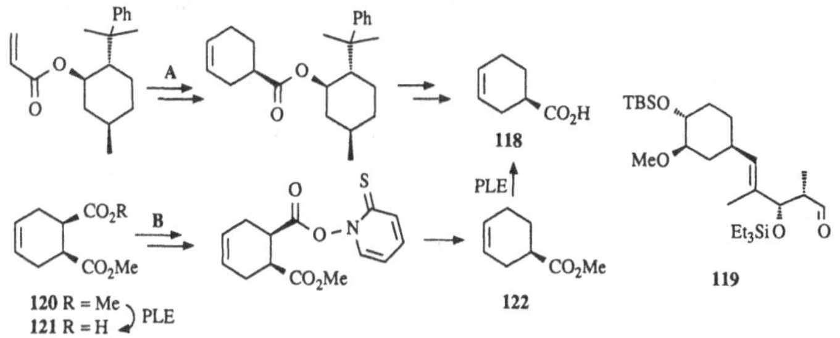

Scheme 52

Part II

The development, mechanistic aspects and synthetic applications of the Barton–McCombie deoxygenation reaction has been discussed at length in several reviews, so, as mentioned earlier, the purpose of the following discussion is to provide a summary of the types of applications this extremely useful reaction has recently seen and a compilation of the relevant references. Free radical-initiated chain reactions where an initial attack is on the C=S bond of the suitable thionocarbonyl derivative of a hydroxyl group results in the homolytic cleavage of the C–O bond. This protocol has witnessed various modifications/adaptations over the years. The following can be considered a general classification[6] of the types of applications this Barton radical method has seen: reductive removal of an –OH group, reductive deoxygenation of diols via cyclic thionocarbonyl derivatives, eliminations from β-functionalized thionocarbonyl derivatives, carbon–carbon bond formation reactions and variations of the initiators or chain transfer agents as alternate methods to generate carbon radicals.

Reductive deoxygenation

$$(R_1)(R_2)\mathrm{CH{-}OH} \longrightarrow (R_1)(R_2)\mathrm{CH{-}H}$$

The deoxygenation of secondary alcohols is a very wide-ranging reaction and is compatible with many functional groups encountered in natural and unnatural product syntheses. The following list of recent applications is demonstrative of this fact: Prelog–Djerassi lactone, the corresponding lactonic acid and C_{19}–C_{27} fragment of rifamycin B[92a], in the synthetic studies on delesserine[92b], the work on acyclic stereocontrol in hetero-conjugate addition and pyranosyl hetero olefin synthesis[93], synthetic studies on decarboxy quadrone[94] (a novel octahydro-3a-7-ethano-3a *H*-Indene skeleton), *S*-hydroxy-ethyl-γ-butyrolactone[95] and R-(+)-γ-caprolactone[95] (chiral building blocks from carbohydrate).

Synthetic studies on sesquiterpenoid phytoalexins[96] ((+)-5-*epi*-aristolochene and (+)-(3*R*)-1-deoxycapsidiol), tetracyclic and pentacyclic quassinoids[97,98] ((+)-picrasin B, (+)-Δ^2-picrasin B, (+)-quassin), (−)-retigeranic acid[99,100] (via [2+3] and [3+4] annulation of enones), synthetic studies on polyene macrolide antibiotics and amphotericin *B*[101] (the 1,3-polyol segments[101]), the C_1-C_{10} fragment[102] of nystatin A_1, isosiccanin methyl ether[103] (a *cis*-fused drimane sesquiterpenoid antibiotic), macrocyclic tricothecene mycotoxins 3-*iso*-verrucarin[104] and verrucinol[105] (trichodermin, calonectrin, deoxynivalenol, etc. fragments), tetracyclic diterpene derivatives[106], an aphidicholin derivative[107] ((+)-aphidicol-15-ene), capnellane family of sesquiterpenoids[108] ((±)-$\Delta^{9(12)}$-capnellene[109]), monoterpene pyridine alkaloid (deoxyrhexifoline[110]), lactarane class of sesquiterpenes (furanether B[111]), (+)-phyllanthocin[112] (an aglycone of the antineoplastic agent phyllanthoside), indolizidine class of alkaloids[113] ((±)-gephyrotoxin *223 AB*), ineupatoriol[114] (a thiophene analog of the potent fish poison ichthyothereol), C_{19}-diterpenoid alkaloids[115] (1-deoxydelphisine, 1-deoxydelcosine, 3-deoxy-aconitine, crassicauline *A*, etc.), prostaglandins PGE_2[116] (a short synthesis via 3-component coupling), isocarbacyclin[117] (a prostacyclin analog which exhibits powerful hypertensive and platelet aggregation properties), ginkgolide *A*[118] (a potent antagonist and insect antifeedent), tetrahydrodicranenone *B*[119] (an antimicrobial fatty acid), pyrrolizidine alkaloids ((−)-supinidine[120], retronecine[120]), (−)-(*R*)-muscone[121], *exo*-brevicomin[122] (a natural pheromone), pinguisone and deoxypinguisone[123] (pinguisane type sesquiterpenes), (+)-averufin[124] (a key intermediate in the aflatoxin biosynthetic pathway), alkaloid matrine[125], descarboxyquadrone[126], heterocyclic pheromones[127] (1,8-dimethyl-3-ethyl-2,9-dioxabicyclo-[3,3,1]-non-7-ene), isocelorbicol[128], 12-(*S*)-hydroxyicosatetrataenoic acid[129] (12-*HETE*), structurally modified glycal (in connection with the synthesis of forskolin) were prepared[130], antibiotic (+)-negamycin[131], (+)-ipomeamarone[132] (a β-substituted furan derivative, a precursor for tenpolin, etc. natural products), (−)-talaromycins *A* and *B*[133], anthracycline antibiotic epirubicin[134], (+)-anamarin[135] (a 6-substituted dihydropyran class of natural product), γ-amino acid statine[136] (an unusual component of general aspartyl protease inhibitor pepstatin), *R*-(+)-α-lipoic acid[137], differentially protected polypropionate subunit polyols[138] (as chiral pool precursors), Ben Derivatives – the γ-turn templates[139] (amino acid derivatives that stabilize secondary structures of polypeptides), hexopyranosides at C-6 and the ensuing stereochemistry of this off-template center was applied in the synthetic studies of macrolide antibiotics,[140] during the study of unusually efficient non-chelate-enforced chirality transfer in electrophilic addition to chiral enolates of 3-deoxy-D-ribofuranose[141], 2′-deoxy-β-disaccharides[142], 4-deoxy derivatives of D-glucopyranose[143] (utilized in the synthetic studies on oligobiosaminide and oligostatin), studies on the thiazole route to higher carbohydrates[144] (via stereoselective homologation of polyalkoxy aldehydes), 2-deoxy-β-D-galactoside[145], 2-deoxy-β-D-glycoside linkages[146,147] (synthetic precursors for antitumor agent mithramycin).

The search for highly stereoselective and high-yielding syntheses of glycosides and oligosaccharides is one of the classical topics of carbohydrate chemistry. The deoxyglycosides are important as components of various natural products of biological significance. During its transport in plants, sucrose is specifically bonded to various proteins. In order to determine the relative importance of interaction of each hydroxyl group of the sugar moiety, a high yielding and yet general procedure for the synthesis of mono (or dideoxy) sugars is necessary, and numerous methods have been developed over the years;

however, a very large number of chemists (judging from the literature) have undoubtedly favored the Barton–McCombie deoxygenation for this transformation.[2-7] Successful applications of this reaction to obtain deoxygenated mono, di and oligosaccharides at a variety of positions have been achieved.[148] Robins and coworkers first adapted Barton deoxygenation to prepare deoxynucleosides in better yields.[149] Their approach has been developed further to prepare various derivatives (azido, fluoro, etc.) of deoxynucleosides and deoxynucleotides which are important precursors for anti-HIV drugs.[150,151]

Cyclic thionocarbonyl derivatives

S
O O
R_1 R_2
→
OH
R_1 R_2
+
OH
R_1 R_2

Upon consideration of the fact that the deoxygenation of primary and secondary alcohols takes place with different efficiency, it was originally envisaged that cyclic thiocarbonates formed from primary and secondary –OH groups would allow selective deoxygenation of the secondary hydroxyl moiety.[152] The reaction has been extended for the convenient preparation of various substrates including nucleosides.[153] Selectivity of deoxygenation of two secondary hydroxyl groups and the scope/mechanisms have been systematically investigated.[154]

Olefination/eliminations

Lythgoe first observed that certain groups which form stabilized free radicals when substituted β to the thionocarbonyl derivative of a hydroxyl moiety undergo smooth elimination upon treatment with trialkyltin radicals to give the olefin.[155] This important observation charted a course for a series of further studies to develop this reaction as a very promising alternative to other olefination reactions.[156] Reaction of a β-epoxy thiocarbonyl ester with tributyltin hydride was first reported by us as an alternative to the Wharton reaction.[157] The reaction has been further studied and developed into an alternative for the preparation of a variety of cyclic and acyclic ethers by Murphy and others.[158]

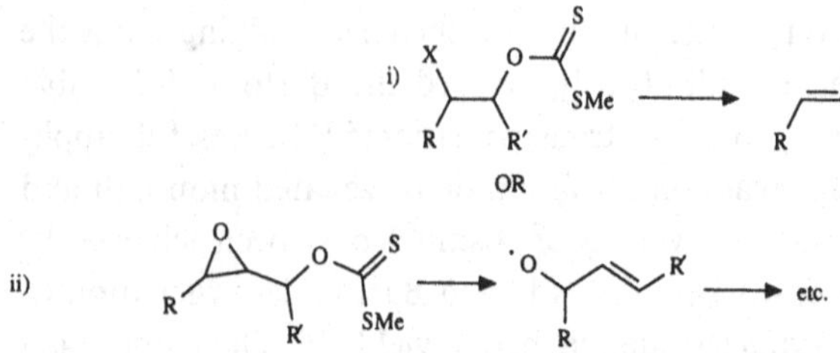

Carbon–carbon bond formation

Thionocarbonyl esters as radical sources in C–C bond forming

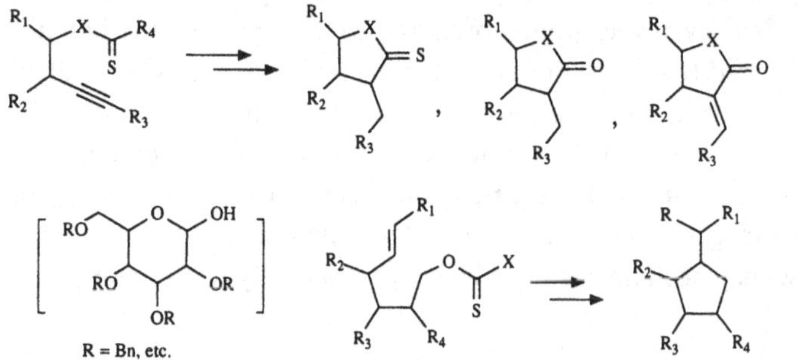

reactions have been very successfully employed in an intramolecular fashion both before and after the fragmentation of the initial adduct. The synthetic applications of this protocol have been reviewed periodically.[2-7] The work of Clive,[159] Bachi[160] and Rajanbabu[161] is especially noteworthy here. Bachi and colleagues have developed this approach into a general method for the synthesis of δ-lactones.[160] Rajanbabu's work in the area of carbohydrates to carbocycles has provided easy access to a variety of chiral precursors for application in the natural product synthesis. A series of elegant publications in this area describe the utilization of Barton deoxygenation protocol to generate carbon radicals. The factors controlling the stereochemical outcome in the corresponding intramolecular addition/cyclization sequences have also been systematically investigated.[161] Recently, Wilcox has also studied the 'carba' analogues of carbohydrates where anomeric oxygen was replaced with a carbon atom, and radical deoxygenation followed by cyclization was the key step in their synthesis.[162] Though there are not many examples of intermolecular trapping of the carbon radicals generated via this method in C–C bond forming reactions, Giese, Keck, Araki and others have demonstrated the principle successfully.[163]

Adaptations/modifications

Although tributyltin- and triphenyltinhydrides have been used for more than thirty years in the dehalogenation and deoxygenation of many types of organic compounds by a radical mechanism, they have, in fact, certain disadvantages for application on a large scale. Tin residues are always formed and are difficult to remove. Organotin compounds are toxic, and the high molecular weights associated with the corresponding H-atom sources make them undesirable for practical reasons.

Our own fruitful efforts (Ph_2SiH_2, Et_3SiH, etc.) in search for other elements in the periodic table which would have weak M–H bonds but strong M–O and M–halogen bonds have been discussed earlier. Chatgilialoglu, Griller and colleagues have also very successfully employed tristrimethylsilylsilane [$(Me_3Si)_3SiH$] with a Si–H bond strength of 79 kcal/mol, comparable with the Sn–H bond strength of 74 ± 2 kcal/mol.[164]

This substitutes very well for most tin hydrides in many reactions. Roberts, on the other hand, has demonstrated that thiols act as polarity reversal catalysts for hydrogen-atom transfer from organosilanes to alkyl radicals.[165] The use of triethyl borane and oxygen as radical initiators[166] has allowed the reaction to be carried out at much lower temperatures and hence mechanistic elucidations have been possible.[167]

Zard and colleagues have developed an interesting radical chain reaction based on the thionocarbonyl derivatives, where S-alkyl, S-acyl and S-alkoxyacyl xanthates are employed as new and very useful sources of alkyl and acyl radicals, as demonstrated in Scheme 53.[168] The alternate undesirable pathways available to an alkyl or alkoxyacyl radical are not important in this reaction due to the reversible and degenerate (path A) nature of the adduct radicals.[168] This reaction does not therefore complete with the expulsion of carbon dioxide (path B), in sharp contrast to previous processes based on stannane

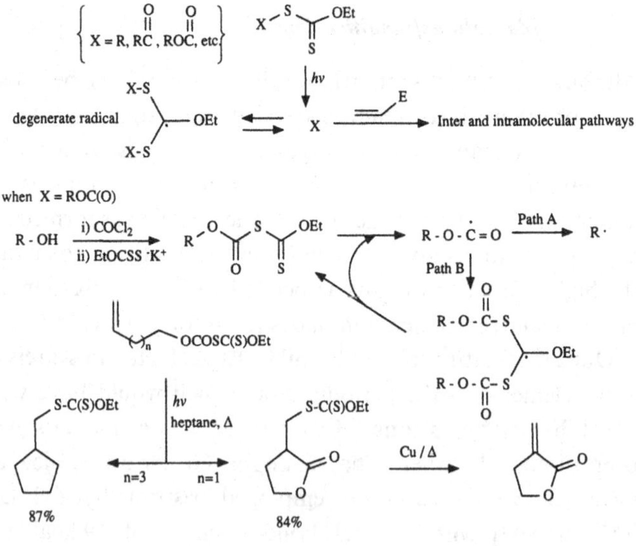

Scheme 53

chemistry or even thiohydroxamate esters (under certain conditions) as described in the previous chapter. Similarly, Minisci and coworkers have also developed a new source of alkyl radicals useful for the selective substitution of heteroaromatic bases.[169] Since the aromatic substitution in this case is an oxidative alkylation, the normal source in the Barton–McCombie protocol was found to be not suitable. Their procedure is very simple and is depicted in Scheme 54.

$$\text{Het-H} + \text{R-O-C}(=\text{S})\text{SMe} + (\text{PhCOO})_2 \longrightarrow \text{Het-R} + \text{PhSC(O)SMe} + \text{PhCO}_2\text{H} + \text{CO}_2$$

Scheme 54

In conclusion it would be appropriate to note that the radical reactions based on the thionocarbonyl functionality offer a plethora of efficient synthetic tools to an organic chemist.

References

1. a) Walling, C. *J. Chem. Educ.* (1986), **63**, 99. b) Mayo, F. R. J. *Chem. Educ.* (1986), **63**, 97. c) Walling, C. *Tetrahedron* (1985), **41**, 3887. d) Beckwith, A. L. J. & Ingold, K. U. In *Rearrangements in Ground and Excited States*, de Mayo, P. (ed.), vol. 1, Academic Press, New York, 1980. e) Minisci, F. & Citterio, A. In *Advances in Free Radical Chemistry*, Williams, G. H. (ed.), vol. 6, Heyden, London, 1980. f) Kochi, J. K. In *Free Radicals*, John Wiley & Sons, New York, 1973. g) Kuivila, H. G. *Acc. Chem. Res.* (1968) **1**, 299. h) Sosnovsky, G. In *Free Radical Reactions in Preparative Organic Chemistry*, MacMillan, New York, 1964.
2. a) Giese, B. *Angew. Chem. Int. Ed. Engl.* (1983), **22**, 753. b) Giese, B. *Angew. Chem. Int. Ed. Engl.* (1985), **24**, 553. c) Giese, B. In *Radicals in Organic Synthesis: Formation of Carbon–Carbon Bonds*; Pergamon: Oxford, 1986.
3. Ramaiah, M. *Tetrahedron* (1987), **43**, 3541.
4. a) Jasperse, C. P., Curran, D. P. & Fevig, T. L. *Chem. Rev.* (1991), **91**, 1237. b) Curran, D. P. *Synthesis* (1988), 417 and 489.

5. Hart, D. J. *Science* (1984), **223**, 883.
6. Crich, D. & Quintero, L. *Chem. Rev.* (1989), **89**, 1413.
7. a) Hartwig, W. *Tetrahedron* (1983), **39**, 2609. b) Hauser, F. M. & Ellenberger, S. R. *Chem. Rev.* (1986), **86**, 35.
8. Cochrane, E. J., Lazer, S. W., Pinhey, J. T. & Whitby, J. D. *Tetrahedron Lett.* (1989), **30**, 7111.
9. a) Winkler, J. D., Hey, J. P & Williard, P. G. *J. Am. Chem. Soc.* (1986), **108**, 6425. b) Winkler, J. D., Henegar, K. E. & Williard, P. G. *J. Am. Chem. Soc.* (1987), **109**, 2850. c) Winkler, J. D., Hey, J. P., Hannon, P. & Williard, P. G. *Heterocycles* (1987), **25**, 55. d) Winkler, J. D., Williard, P. G. & Hey, J. P. *Tetrahedron Lett.* (1988), **29**, 4691.
10. Winkler, J. D., Scott, R. D. & Williard, P. G. *J. Am. Chem. Soc.* (1990), **112**, 8971.
11. Winkler, J. D. & Sridar, V. *J. Am. Chem. Soc.* (1986), **108**, 1708.
12. Skuballa, N., Musso, H. & Boland, W. *Tetrahedron Lett.* (1990), **31**, 497.
13. Liu, Z. J. & Xu, F. *Tetrahedron Lett.* (1989), **30**, 3457.
14. Otterbach, A. & Musso, H. *Angew. Chem. Int. Ed. Engl.* (1987), **26**, 554.
15. a) Ihara, M., Kametani, T., Kabuto, C., Fukumoto, K. & Suzuki, M. *J. Am. Chem. Soc.* (1988), **110**, 1963. b) Ihara, M., Suzuki, M., Fukumoto, K. & Kakuto, C. *J. Am. Chem. Soc.* (1990), **112**, 1164.
16. Michl, J. & Gladysz, J. A. *Chem. Rev.* (1989), **89**, 973.
17. Hedberg, L., Hedberg, K., Eaton, P. E., Nordai, N. & Robiette, A. G. *J: Am. Chem. Soc.* (1991), **113**, 1514.
18. Castaldi, G., Colombo, R. & Allegrini, P. *Tetrahedron Lett.* (1991), **32**, 2173.
19. a) Della, E. W. & Tsanaktsidis, J. *Aust. J. of Chem.* (1986), **39**, 2061. b) Della, E. W. & Tsanaktsidis, J. *Aust. J. of Chem.* (1989), **42**, 61.
20. Kaszynski, P., McMurdie, N. D. & Michl, J. *J. Org. Chem.* (1991), **56**, 307.
21. Sternbach, D. D. & Eusinger, C. L. *J. Org. Chem.* (1990), **55**, 2725.
22. Magnus, P., Ludlow, M., Kim, C. S. & Buniface, P. *Heterocycles* (1989), **28**, 951.

23. Grieco, P. A. & Abood, N. *J. Chem. Soc., Chem. Commun.* (1990), 410.
24. Compopiano, O., Little, R. D. & Petersen, J. L. *J. Am. Chem. Soc.* (1985), **107**, 3721.
25. Braekman, J. C., Daloze, D., Kaisin, M. & Moussiaux, B. *Tetrahedron* (1985), **41**, 4603.
26. Linz, G., Weetman, J., Abdel Hady, A. F. & Helmchen, G. *Tetrahedron Lett.* (1989), **30**, 5599.
27. Sardina, F. J., Howard, M. H., Morningstar, M. & Rapoport, H. *J. Org. Chem.* (1990), **55**, 5025.
28. a) Geise, B. *Angew. Chem. Int. Ed. Engl.* (1989), **28**, 969 and references cited therein. b) Rajanbabu, T. V. *Acc. Chem. Res.* (1991), **24**, 139.
29. a) Use of *tert*-dodecyl mercaptan has been recommended by Crich because it is easier to work with compared to *tert*-butyl thiol. b) Crich, D. & Ritchie, T. J. *J. Chem. Soc., Chem. Commun.* (1988), 1461. c) Crich, D. & Lim, L. B. L. *Tetrahedron Lett.* (1990), **31**, 1897. d) Crich, D. & Ritchie, T. J. *Carbohydrate Res.* (1989), **190**, C3.
30. a) Dauben, W. G., Kowalcz, B. A. & Bridon, D. P. *Tetrahedron Lett.* (1989), **30**, 2461. b) Stofer, E. & Lion, C. *Bull. Soc. Chim. Belg.* (1987), **96**, 623.
31. Adcock, W. & Gangodawile, H. *J. Org. Chem.* (1989), **54**, 6040.
32. Vogel, E., Lex, J., Schieb, T., Schmickler, H., Schmidt, K. & Schulz, W. H. *Angew. Chem. Intl. Ed. Engl.* (1986), **25**, 723.
33. a) Harvey, R. G., Pataki, J., Cortez, C., Diraddo, P. & Yang, C. X. *J. Org. Chem.* (1991), **56**, 1210. b) Rösslein, L. & Tamm, C. *Helv. Chim. Acta* (1988), **71**, 47.
34. a) Fleet, G. W. J., Son, J. C., Peach, J. M. & Hamor, T. S. *Tetrahedron Lett.* (1988), **29**, 1449. b) Fleet, G. W. J., Son, J. C., Vogt, K., Peach, J. M. & Hamor, T. S. *Tetrahedron Lett.* (1988), **29**, 1451.
35. a) Witty, D. R., Fleet, G. W. J., Vogt, K., Wilson, F. X., Wang, Y., Storer, R., Myers, P. L. & Wallis, C. J. *Tetrahedron Lett.* (1990), **31**, 4787. b) Wilson, F. X., Fleet, G. W. J., Vogt, K., Wang, Y., Witty, D. R., Chui, S. & Storer, R. *Tetrahedron Lett.* (1990), **31**, 6931. c) Wang, Y., Fleet, G. W. J., Wilson, C. J.,

Storer, R., Myers, P. L, Wallis, C. J., Doherty, O., Watkin, D. J., Peach, J. M. *Tetrahedron Lett.* (1991), **32**, 1675. d) Wang, Y., Fleet, G. W. J., Storer, R., Myers, P. L., Wallis, C. J. & Doherty, O. *Tetrahedron Asymm.* (1990), **1**, 527.
36. Wenkert, E., Arrhenius, T. S., Bookser, B., Guo, M. & Mancini, P. *J. Org. Chem.* (1990), **55**, 1185.
37. Kobayashi, S., Kamiyam, K. & Ohno, M. *J. Org. Chem.* (1990), **55**, 1169.
38. Tsanaktsidis, J. & Eaton, P. E. *Tetrahedron Lett.* (1989), **30**, 6967.
39. Schäfer, J. & Szeimies, G. *Tetrahedron Lett.* (1990), **31**, 2263.
40. Drost, K. J. & Cava, M. P. *J. Org. Chem.* (1991), **56**, 2240.
41. a) Bloodworth, A. J., Crich, D. & Melvin, T. *J. Chem. Soc., Chem. Commun.* (1987), 786. b) Bloodworth, A. J., Crich, D. & Melvin, J. *J. Chem. Soc., Perkin Trans. I* (1990), 2957.
42. Biovin, J., Crépon, E. & Zard, S. Z. *Tetrahedron Lett.* (1990), **31**, 6869.
43. Boivin, J., da Silva, E., Ourisson, G. & Zard, S. Z. *Tetrahedron Lett.* (1990), **31**, 2501.
44. Kawamura, T., Kaoma, T. & Yonezawa, T. *J. Am. Chem. Soc.* (1970), **92**, 7222.
45. Reich, I. L. & Reich, H. J. *J. Org. Chem.* (1990), **55**, 2282.
46. a) Curran, D. P. & van Elburg, P. A. *Tetrahedron Lett.* (1989), **30**, 2501. b) Curran, D. P., Chen, M.-H., Spletzer, E., Seong, C. M. & Chang, C.-T. *J. Am. Chem. Soc.* (1989), **111**, 8872.
47. a) Feldman, K. S., Ramanelli, A. L., Ruckle, R. E. & Miller, R. F. *J. Am. Chem. Soc.* (1988), **110**, 3300. b) Feldman, K. S. & Vong, A. K. K. *Tetrahedron Lett.* (1990), **31**, 823.
48. a) Clive, D. L. J. & Anogh, A. C. *J. Chem. Soc., Chem. Commun.* (1985), 980. b) Miura, K., Fugami, K., Oshima, K. & Utimoto, K. *Tetrahedron Lett.* (1988), **29**, 5135.
49. a) Čeković, Ž. & Saičić, R. *Tetrahedron Lett.* (1986), **27**, 5893. b) Čeković, Ž. & Saičić, R. *Tetrahedron Lett.* (1990), **31**, 4203. c) Čeković, Ž. & Saičić, R. *Tetrahedron Lett.* (1990), **31**, 6085.
50. Amad Junan, S. A., Walkington, A. J. & Whiting, D. A. *J. Chem. Soc., Chem. Commun.* (1989), 1613.
51. Walkington, A. J. & Whiting, D. A. *Tetrahedron Lett.* (1989), **30**, 4731.
52. Crombie, L. *Nat. Prod. Rep.* (1984), **1**, 3.

53. a) Togo, H., Fujii, M. & Yokoyama, M. *J. Synth. Org. Chem. Jpn.* (1990), **48**, 641. b) Togo, H., Fujii, M., Ikuma, T. & Yokoyama, M. *Tetrahedron Lett.* (1991), **32**, 3377.
54. a) Giese, B. *Angew. Chem., Int. Ed. Engl.* (1989), **28**, 967. b) Rajanbabu, T. V. *Acc. Chem. Res.* (1991), **24**, 139. c) Barton, D. H. R. in references 76–80 of Chapter 5.
55. a) Porter, N. A., Swann, E., Nally, J. & McPhail, A. T. *J. Am. Chem. Soc.* (1990), **112**, 6740. b) Scott, D. M., McPhail, A. T. & Porter, N. A. *Tetrahedron Lett.* (1990), **31**, 1679. c) Porter, N. A., Breyer, R., Swann, E., Nally, J., Pradhan, J., Allen, T. & McPhail, A. T. *J. Am. Chem. Soc.* (1991), **111**, 7002.
56. Giese, B., Zehnder, M., Roth, M. & Zeitz, H.-G. *J. Am. Chem. Soc.* (1990), **112**, 6741.
57. Crich, D. & Davies, J. W. *Tetrahedron Lett.* (1987), **28**, 4205.
58. Beckwith, A. L. J. & Chai, C. L. L. *J. Chem. Soc., Chem. Commun.* (1990), 1087.
59. Gawronska, K., Gawronski, J. & Walborsky, H. M. *J. Org. Chem.* (1991), **56**, 2193.
60. Boche, G. & Walborsky, H. M. In *Cyclopropane derived Reactive Intermediates*, John Wiley & Sons, New York, 1990.
61. Brandi, A., Cicchi, S., Goti, A. & Pietrusiewicz, K. *Tetrahedron Lett.* (1991), **32**, 3265.
62. a) Newcomb, M., Kumar, M. U., Boivin, J., Crepon, E. & Zard, S. Z. *Tetrahedron Lett.* (1991), **32**, 45. b) Beckwith, A. L. J. & Davidson, I. G. E. *Tetrahedron Lett.* (1991), **32**, 49.
63. a) Ovchinnikov, A. A. *Theoret. Chim. Acta* (1978), **47**, 297. b) Togo, Y., Nakamura, N. & Iwamura, H. *Chem. Lett.* (1991), 1201.
64. Bergbreiter, D. E. & Zhou, J. *J. Poly. Sc. Part A. Poly. Chem.* (1992), **30**, 2049.
65. Beckwith, A. L. J. & Hay, B. P. *J. Am. Chem. Soc.* (1988), **110**, 4415.
66. Newcomb, M. & Deeb, T. M. *J. Am. Chem. Soc.* (1987), **109**, 3163.
67. Newcomb, M. & Marguardt, D. J. *Heterocycles* (1989), **28**, 129.
68. Newcomb, M., Marguardt, D. J. & Kumar, M. U. *Tetrahedron* (1990), **46**, 2545.
69. Newcomb, M., Deeb, T. M. & Marguardt, D. J. *Tetrahedron* (1990), **46**, 2317.

70. Newcomb, M., Marguardt, D. J. & Deeb, T. M. *Tetrahedron* (1990), **46**, 2329.
71. Newcomb, M. & Kumar, M. U. *Tetrahedron Lett.* (1990), **31**, 1675.
72. Newcomb, M. & Weber, K. A. *J. Org. Chem.* (1991), **56**, 1309.
73. Newcomb, M. & Esker, J. L. *Tetrahedron Lett.* (1991), **32**, 1035.
74. a) Togo, H. & Yokoyama, M. *Heterocycles* (1990), **31**, 437. b) Togo, H., Fujii, M. & Yokoyama, M. *Bull. Chem. Soc. Jpn.* (1991), **64**, 57.
75. Crich, D. & Fortt, S. M. *Synthesis* (1987), 35.
76. a) Kochi, J. K., Sheldon, R. A. *Metal Catalysed Oxidations of Organic Compounds*, Academic Press New York, 1981. b) Walling, C. *Acc. Chem. Res.* (1975), **8**, 125 and references cited therein.
77. Boivin, J., Fouquet, E. & Zard, S. Z. *Tetrahedron Lett.* (1991), **32**, 4299.
78. a) Boivin, J., Fouquet, E. & Zard, S. Z. *Tetrahedron Lett.* (1990), **31**, 85. b) Boivin, J., Fouquet, E. & Zard, S. Z. *Tetrahedron Lett.* (1991), **31**, 3545.
79. Boivin, J., Fouquet, E. & Zard, S. Z. *J. Am. Chem. Soc.* (1991), **113**, 1054.
80. a) Avila, L. Z., Loo, S. H. & Frost, J. W. *J. Am. Chem. Soc.* (1987), **109**, 6758. b) Shames, S. L., Wackett, L. P., LaBarge, M. S. & Kuczkowski, R. L. *Bioorg. Chem.* (1987), **15**, 366.
81. Avila, L. Z. & Frost, J. W. *J. Am. Chem. Soc.* (1988), **110**, 7904.
82. a) Dauben, W. G., Bridon, D. P. & Kowalczyk, B. A. *J. Org. Chem.* (1989), **54**, 6101. b) Dauben, W. G., Bridon, D. P. & Kowalczyk, B. A. *J. Org. Chem.* (1990), **55**, 376.
83. Newcomb, M. & Park, S. U. *J. Am. Chem. Soc.* (1986), **108**, 4132.
84. a) Newcomb, M. & Kaplan, J. *Tetrahedron Lett.* (1987), **28**, 1615. b) Newcomb, M. & Kaplan, J. *Tetrahedron Lett.* (1988), **29**, 3449. c) Park, S.-U., Varick, T. R. & Newcomb, M. *Tetrahedron Lett.* (1990), **31**, 2975. d) Newcomb, M., Manek, M. B. & Glenn, A. G. *J. Am. Chem. Soc.* (1991), **113**, 949. e) Curran, D. P., Bosch, E., Kaplan, J. & Newcomb, M. *J. Org. Chem.* (1989), **54**, 1826.
85. Lusztyk, J., Maillard, B., Deycard, S., Lindsay, D. A. & Ingold,

K. U. *J. Org. Chem.* (1987), **52**, 3509.

86. Bohne, C., Boch, R. & Scaiano, J. C. *J. Org. Chem.* (1990), **55**, 5414.
87. Beckwith, A. L. J. & Hay, B. P. *J. Am. Chem. Soc.* (1989), **111**, 230.
88. a) Beckwith, A. L. J. & Zimmerman, J. *J. Org. Chem.* (1991), **56**, 5791. b) Beckwith, A. L. J., Bowry, V. W. & Schiesser, C. H. *Tetrahedron* (1991), **47**, 121.
89. Newcomb, M. & Glenn, A. G. *J. Am. Chem. Soc.* (1989), **111**, 275.
90. Newcomb, M. & Manek, M. B. *J. Am. Chem. Soc.* (1990), **112**, 9662.
91. Kocieński, P., Stocks, M., Donald, D. & Perry, M. *SynLett* (1990), 38.
92. a) Ziegler, F. E., Cain, W. T., Kneisley, A., Stirchak, E. P. & Wester, R. T. *J. Am. Chem. Soc.* (1988), **110**, 5442. b) Kinoshita, T., Yoshida, N. & Miwa, T. *Bull. Chem. Soc. Jpn.* (1990), **63**, 1538.
93. Isobe, M., Ichikawa, Y., Funabashi, Y., Mio, S. & Goto, T. *Tetrahedron* (1986), **42**, 2863.
94. Imanishi, T., Matsui, M., Yamashita, M. & Iwata, C. *Tetrahedron Lett.* (1986), **27**, 3161.
95. Gurjar, M. K. & Patil, V. J. *Indian J. Chem.* (1986), **25B**, 596.
96. Whitehead, I. M., Ewing, D. F., Threlfall, D. R., Cane, D. E. & Prabhakaran, P. C. *Phytochem.* (1990), **29**, 479.
97. Kim, M., Kawada, K., Gross, R. S. & Watt, D. S. *J. Org. Chem.* (1990), **55**, 504.
98. Kawada, K., Kim, M. & Watt, D. S. *Tetrahedron Lett.* (1989), **30**, 5989.
99. Hudlicky, T., Fleming, A. & Radesca, L. *J. Am. Chem. Soc.* (1989), **111**, 6691.
100. Hudlicky, T., Radesca-Kwart, L., Li, L.-G. & Bryant, T. *Tetrahedron Lett.* (1988), **29**, 3283.
101. a) Nakata, T., Suenaga, T. & Oishi, T. *Tetrahedron Lett.* (1989), **30**, 6525. b) Liang, D., Pauls, H. W., Fraser-Reid, B., Georges, M., Mubarak, A. M. & Jarosz, S. *Can. J. Chem.* (1986), **64**, 1800. c) Oppong, I., Pauls, H. W., Liang, D. & Fraser-Reid, B. *J. Chem. Soc., Chem. Commun.* (1986), 1241.

102. Prandi, J. & Beau, J.-M. *Tetrahedron Lett.* (1989), **30**, 4517.
103. Liu, H.-J. & Ramani, B. *Tetrahedron Lett.* (1988), **29**, 6721.
104. Tamm, C. & Jeker, N. *Tetrahedron* (1989), **45**, 2385.
105. Banerjee, A. K., Acevedo, J. C. & Canudas-González, N. *Bull. Soc. Chim. Belg.* (1990), **99**, 9.
106. Kelly, R. B., Lal, G. S., Gowda, G. & Rej, R. N. *Can. J. Chem.* (1984), **62**, 1930.
107. Mehta, G. K., Murthy, A. N., Reddy, D. S. & Reddy, A. V. *J. Am. Chem. Soc.* (1986), **108**, 3443.
108. Piers, E. & Karunaratne, V. *Tetrahedron* (1989), **45**, 1089.
109. Ranarivelo, Y., Hotellier, F., Skaltsounis, A.-L. & Tillequin, F. *Heterocycles* (1990), **31**, 1727.
110. Price, M. E. & Schore, N. E. *J. Org. Chem.* (1989), **54**, 5662.
111. Martin, S. F., Dappen, M. S., Durpé, B., Murphy, C. J. & Colapret, J. A. *J. Org. Chem.* (1989), **54**, 2209.
112. Brandi, A., Cordero, F. & Querci, C. *J. Org. Chem.* (1989), **54**, 1748.
113. Still, I. W. J., Banait, N. S. & Frazer, D. V. *Synth. Commun.* (1988), **18**, 1461.
114. Kulanthaivel, P. & Pelletier, S. W. *Tetrahedron* (1988), **44**, 4313.
115. Suzuki, M., Kawagishi, T., Yarragisawa, A., Suzuki, T., Okamura, N. & Noyori, R. *Bull. Chem. Soc., Jpn.* (1988), **61**, 1299.
116. Suzuki, M., Koyano, H. & Noyori, R. *J. Org. Chem.* (1987), **52**, 5583.
117. Corey, E. J. & Ghosh, A. K. *Tetrahedron Lett.* (1988), **29**, 3205.
118. a) Moody, C. J., Roberts, S. M. & Toczek, J. *J. Chem. Soc., Chem. Commun.* (1986), 1292. b) Moody, C. J., Roberts, S. M. & Toczek, J. *Chem. Soc., Perkin Trans. 1* (1988), 1401.
119. Gruszecka-Kowalik, E. & Zalkow, L. H. *J. Org. Chem.* (1990), **55**, 3398.
120. Gurjar, M. K., Patil, V. J. & Pawar, S. M. *Indian J. Chem.* (1987), **26B**, 1115.
121. Porter, N. A., Lacher, B., Chang, V. H.-T. & Magnin, D. R. *J. Am. Chem. Soc.* (1989), **111**, 8309.
122. Yadav, J. S., Reddy, P. S. & Joshi, B. V. *Tetrahedron Lett.* (1988), **44**, 7243.
123. Uyehara, T., Kabasawa, Y. & Kato, T. *Bull. Chem. Soc. Jpn.*

(1986), **59**, 2521.
124. Koreeda, M., Hulin, B., Yoshihara, M., Townsend, C. A. & Christensen, S. B. *J. Org. Chem.* (1985), **50**, 5426.
125. Chen, J., Browne, L. J. & Gonnela, N. C. *J. Chem. Soc., Chem. Commun.* (1986), 905.
126. Kakiuchi, K., Ue, M., Tadaki, T. & Tobe, Y. *Chem. Lett.* (1986), 507.
127. Mori, K. & Kisida, H. *Tetrahedron Lett.* (1986), **42**, 5281.
128. In this unusual and difficult to synthesize polyhydroxylated derivative of agarofuran there are six axial substituents. Barton protocol was successfully employed at the later stages of the synthesis. Huffman, J. W. & Raveendranath, P. C. *J. Org. Chem.* (1986), **51**, 2148.
129. Manna, S., Viala, J., Yadagiri, P. & Falck, J. R. *Tetrahedron Lett.* (1986), **27**, 2679.
130. Paquette, L. A. & Oplinger, J. A. *J. Org. Chem.* (1988), **53**, 2953.
131. De Bernardo, S., Tengi, J. P., Sasso, G. & Weigele, M. *Tetrahedron Lett.* (1988), **29**, 4077.
132. Marco, J. L. *Tetrahedron* (1989), **45**, 1475.
133. Cubero, I. I., Plaza, M. T. & López-Espinosa, P. *Carbohydrate Res.* (1990), **205**, 293.
134. Florent, J.-C., Ughetto-Monfrin, J. & Monneret, C. *Carbohydrate Res.* (1988), **181**, 253.
135. Gillard, F., Heissler, D. & Riehl, J.-J. *J. Chem Soc., Chem. Commun.* (1988), 229.
136. Yanagisawa, H., Kanazaki, T. & Nishi, T. *Chem. Lett.* (1989), 687.
137. Rama Rao, A. V., Gurjar, M. K., Garyali, K. & Ravindranathan, T. *Carbohydrate Res.* (1986), **148**, 51.
138. Marshall, J. A. & Blough, B. E. *J. Org. Chem.* (1990), **55**, 1540.
139. Kemp, D. S. & Carter, J. S. *J. Org. Chem.* (1989), **54**, 109.
140. Jones, K. & Wood, W. W. *J. Chem. Soc., Perkin Trans. 1* (1988), 999.
141. Mulzer, J., Steffen, U., Zorn, L., Schneider, C., Weinhold, E., Münch, W., Rudert, R., Luger, P. & Hartl, H. *J. Am. Chem. Soc.* (1988), **110**, 4640.
142. Trumtel, M., Veyrières, A., & Sinay, P. *Tetrahedron Lett.* (1989), **30**, 2539.

143. Shibata, Y., Kosuge, Y. & Ogawa, S. *Carbohydrate Res.* (1990), **199**, 37.
144. Dondoni, A., Fautin, G., Fogagnola, M., Medici, A. & Pedrini, P. *J. Org. Chem.* (1989), **54**, 693.
145. a) Kihlberg, J., Frejd, T., Jansson, K., Sundin, A. & Magnusson, G. *Carbohydrate Res.* (1988), **176**, 271. b) *Idem*, ibid., (1986), **152**, 113.
146. Gurjar, M. K. & Ghosh, P. K. *Indian J. Chem.* (1988), **27B**, 1063.
147. Trumtel, M., Tavecchia, P., Veyrières, A., Sinay, P. *Carbohydrate Res.* (1989), **191**, 29.
148. For representative examples see, a) Descotes, G., Mentech, J. & Roques, N. *Carbohydrate Res.* (1989), **188**, 63. b) Kihlberg, J., Frejd, T., Jansson, K., Kitzing, S. & Magnusson, G. *Carbohydrate Res.* (1989), **185**, 171. c) Okabe, M., Sun, R.-C. & Zenchoff, G. B. *J. Org. Chem.* (1991) **56**, 4392. d) Meuwly, R. & Vasella, A. *Helv. Chim. Acta* (1986), **69**, 751. e) Binder, T. P. & Robyt, J. F. *Carbohydrate Res.* (1986), **147**, 149. f) Fiander, J. & De Las Heras, F. G. *Carbohydrate Res.* (1986), **153**, 325. g) Huang, J. T., Chen, L. C., Wang, L., Kim, M. H., Warshaw, J. A., Armstrong, D., Zhu, Q. Y., Chou, T. C., Watanabe, K. A., Matulicadamic, J., Su, T. L., Fox, J. J., Polsky, B., Baron, P. A., Gold, J. W. M., Hardy, W. D. & Zuckerman, E. *J. Med. Chem.* (1991), **34**, 1640. h) Blattner, R., Furneaux, R. H., Mason, J. M. & Tyler, P. C. *Pest. Sci.* (1991), **31**, 419. i) Shing, T. K. M. & Tang, Y. *J. Chem. Soc., Chem. Commun.* (1990), 748. j) Lindhorst, T. K. & Thiem, J. *Leib. Ann. Chem.* (1990), 1237. k) Svridov, A. F., Borodkin, V. S., Ermolenko, M. S., Yashunsky, D. V. & Kochetkov, N. K. *Tetrahedron* (1991), **47**, 2291. l) Benhaddou, R., Czernecki, S., Valery, J. M. & Bellosta, V. *Bull. Soc. Chim. Fr.* (1991), 108.
149. a) Robins, M. J. & Wilson, J. S. *J. Am. Chem. Soc.* (1981), **103**, 932. b) Robins, M. J., Wilson, J. S. & Hansske, F. *J. Am. Chem. Soc.* (1983), **105**, 4059.
150. a) Robins, M. J. & Zou, R. *Can. J. Chem.* (1987), **65**, 1436. b) Robins, M. J., Imbach, J. L., Gosselin, G., Hansske, F., Madej, D., Wilson, J. S., Bergogne, M. C., Balzarini, J. & Declerecq, E. *Can. J. Chem.* (1988), **66**, 1258 and references cited therein.

c) Robins, M. J., Zou, R., Madej, D., Hansske, F. & Tyrrel, D. L. J. *Nucleosides Nucleotides* (1989), **8**, 725 and references cited therein. d) Somano, V. & Robins, M. J. *J. Org. Chem.* (1990), **55**, 5186. e) Somano, V. & Robins, M. J. *Synthesis* (1991), 283 and references cited therein.

151. For representative examples see, a) Huang, W. C., Orban, J., Kintanar, A., Reid, B. R. & Drobny, G. P. *J. Am. Chem. Soc.* (1990), **112**, 9059. b) Seela, F. & Bourgeois, W. *Synthesis* (1990), 945. c) Bamford, M. J., Coe, P. L. & Walker, R. T. *J. Med. Chem.* (1990), **33**, 2494. d) Martin, J. A., Bushnell, D. J., Duncan, I. B., Dunsdon, S. J., Hall, M. J., Machin, P. J., Merrett, J. H., Parkes, K. E. B., Roberts, N. A., Thomas, G. J., Galpin, S. A. & Kinchington, D. *J. Med. Chem.* (1990), 2137. e) Wingerinck, P., Vanaerschot, A., Janssen, G., Class, P., Balzarini, J., Declercq, E. & Hardewijn, P. *J. Med. Chem.* (1990), **33**, 868. f) Sekine, M. & Nakanishi, T. *J. Org. Chem.* (1990), **55**, 924. g) Sasaki, M., Murae, T. & Takahashi, T. *J. Org. Chem.* (1990), **55**, 528. h) Lin, T. S., Yang, J. H., Liu, M. C. & Zhu, J. L. *Tetrahedron Lett.* (1990), **31**, 3829. i) Vanaerschot, A., Hardewijn, P., Balzarini, J., Declercq, E. & Pauwels, R. *J. Med. Chem.* (1989), **32**, 1743. j) Seela, F. & Muth, H.-P. *Liebigs Ann. Chem.* (1988), 215. k) Seela, F. & Driller, H. *Helv. Chim. Acta* (1988), **71**, 757. l) Biggadike, K., Borthwick, A. D., Exall, A. M., Kirk, B. E., Roberts, S. M. & Youds, P. *J. Chem. Soc., Chem. Commun.* (1987), 1083. m) Pathak, T. & Chattopadhyaya, J. *Tetrahedron* (1987), **43**, 4227. n) Wu, J. C., Bazin, H. & Chattopadhyaya, J. *Tetrahedron* (1987), **43**, 2255.

152. Barton, D. H. R. & Subramanian, R. *J. Chem. Soc., Perkin Trans. 1* (1977), 1718.

153. a) Suzuki, M., Yanagisawa, A. & Noyori, R. *Tetrahedron Lett.* (1984), **25**, 1383. b) Redlich, H., Sudau, W. & Paulsen, H. *Tetrahedron* (1985), **41**, 4253. c) Alpegiana, M. & Hanessian, S. *J. Org. Chem.* (1987), **52**, 278. d) Giese, B., Groninger, K. S., Witzel, T., Korth, H. G. & Sustmann, R. *Angew. Chem. Int. Ed. Engl.* (1987), **26**, 233. e) De Bernardo, S., Tengi, J. P., Sassoand, F. & Weigele, M. *Tetrahedron Lett.* (1988), **29**, 4077. f) Mills, S., Desmond, R., Reamer, R. A., Volante, R. P. & Shinkai, I.

Tetrahedron Lett. (1988), **29**, 281. g) Beckwith, A. L. J., Davies, A. G., Davison, I. G. E., Maccoll, A. & Mruzek, M. H. *J. Chem. Soc., Chem. Commun.* (1988), 475 and references cited therein.

154. a) Pouzar, S., Vavsickova, S., Dasar, P., Cerny, I. & Haxel, M. *Coll. Czech. Chem. Commun.* (1983), 2423. b) Patroni, J. J., Stick, R. V., Engelhardt, L. M. & White, A. H. *Aust. J. Chem.* (1986), **39**, 699. c) Kanemitsu, K., Tsuda, Y., Hague, M. E., Tsubono, K. & Kikuchi, T. *Chem. Pharm. Bull* (1987), **35**, 3874. d) Kanemitsu, K., Tsuda, Y., Kakimoto, K. & Tohru, K. H. *Chem. Pharm. Bull.* (1987), **35**, 2148. e) Ziegler, F. E. & Zheng, Z.-L. *J. Org. Chem.* (1990), **55**, 1416.
155. Lythgoe, B. & Waterhouse, L. *Tetrahedron Lett.* (1977), 4223.
156. a) Beckwith, A. L. J. & Pigou, P. E. *Aust. J. Chem.* (1986), **39**, 77. b) Beckwith, A. L. J. & Pigou, P. E. *Aust. J. Chem.* (1986), **39**, 1151. c) Oppong, L., Pauls, H. W., Liang, D. & Fraser-Reid, B. *J. Chem. Soc., Chem. Commun.* (1986), 1241. d) Barrish, J. C., Lee, H. L., Mitt, T., Pizzolato, G., Baggiolini, E. G. & Uskovič, J. *J. Org. Chem.* (1988), **53**, 4282.
157. Barton, D. H. R., Motherwell, R. S. H. & Motherwell, W. B. *J. Chem. Soc., Perkin Trans. 1* (1981), 2363.
158. a) Johns, A. & Murphy, J. A. *Tetrahedron Lett.* (1988), **29**, 837. b) Murphy, J. A., Patterson, C. W. & Wooster, N. F. *Tetrahedron Lett.* (1988), **29**, 955. c) Cook, M., Hares, O., Johns, A., Murphy, J. A. & Patterson, C. W. *J. Chem. Soc., Chem. Commun.* (1986), 1419. d) Johns, A., Murphy, J. A. & Patterson, C. W. *J. Chem. Soc., Chem. Commun.* (1987), 1238.
159. Angoh, A. G. & Clive, D. L. J. *J. Chem. Soc., Chem. Commun.* (1985), 980.
160. a) Bachi, M. D. & Bosch, E. *J. Org. Chem.* (1989), **54**, 1234. b) Bachi, M. D. & Bosch, E. *J. Chem. Soc., Perkin Trans. 1* (1988), 1517. c) Bachi, M. D. & Bosch, E. *Tetrahedron Lett.* (1986), **27**, 641.
161. a) Rajanbabu, T. V. *Acc. Chem. Res.* (1991), **24**, 139. b) Rajanbabu, T. V., Fukunaga, T. & Reddy, G. S. *J. Am. Chem. Soc.* (1989), **111**, 1759. c) Rajanbabu, T. V. & Fukunaga, T. *J. Am. Chem. Soc.* (1989), **111**, 296. d) Rajanbabu, T. V. *J. Org. Chem.* (1988), **53**, 4522. e) Rajanbabu, T. V. *J. Am. Chem. Soc.* (1987), **109**, 609.

162. Gaudino, J. J. & Wilcox, C. S. *J. Am. Chem. Soc.* (1990), **112**, 4374.
163. a) Giese, B., Gonzalez-Gomez, J. A. & Witzel, T. *Angew. Chem. Int. Ed. Engl.* (1984), **23**, 69. b) Keck, G. E., Enholm, E. J., Yates, J. B. & Wiley, M. R. *Tetrahedron* (1985), **41**, 4079. c) Araki, Y., Endo, T., Tanji, M., Magasawa, J. & Ishido, Y. *Tetrahedron Lett.* (1987), **28**, 5853. d) Araki, Y., Endo, T., Tanji, M., Magasawa, J. & Ishido, Y. *Tetrahedron Lett.* (1988), **29**, 351.
164. a) Kanabus-Kaminska, J. M., Hawari, J. A., Griller, D. & Chatgilialoglu, C. *J. Am. Chem. Soc.* (1987), **109**, 5267. b) Chatgilialoglu, C., Griller, D. & Lesage, M. *J. Org. Chem.* (1988), **53**, 3641. c) Chatgilialoglu, C., Griller, D. & Lesage, M. *Tetrahedron Lett.* (1989), **30**, 2733. d) Kulicke, K. J. & Giese, B. *Syn. Lett.* (1990), 91. e) Chatgilialoglu, C., Guerrini, A. & Seconi, G. *Tetrahedron Lett.* (1990), 219. f) Lesage, M., Martinho-Simoes, J. A. & Griller, D. *J. Org. Chem.* (1990), **55**, 5413. g) Schummer, D. & Höfle, G. *Syn. Lett.* (1990), 705. h) Ballestri, M., Chatgilialoglu, C., Clark, K. B., Griller, D., Giese, B. & Kopping, B. *J. Org. Chem.* (1991), **56**, 678.
165. Allen, R. P., Roberts, B. P. & Willis, C. R. *J. Chem. Soc., Chem. Commun.* (1989), 1387.
166. Nozaki, K., Oshima, K. & Utimoto, K. *Tetrahedron Lett.* (1988), **29**, 6125.
167. See references 18–21 of Chapter 4.
168. a) Delduc, P., Tailhan, C. & Zard, S. Z. *J. Chem. Soc., Chem. Commun.* (1988), 308. b) Forbes, J. E. & Zard, S. Z. *J. Am. Chem. Soc.* (1990), **112**, 2034.
169. Coppa, F., Fontana, F., Minisci, F., Pianese, G. & Zhao, L. *Tetrahedron Lett.* in press.

Author index

The first number is the page reference, the superscript is the number of the reference at the end of the chapter.

Subject index

Printed in the United States
By Bookmasters